CW01303101

LATHKILL DALE DERBYSHIRE
its Mines and Miners

J. H. Rieuwerts

PEAK DISTRICT MINING MUSEUM, MATLOCK BATH

Founded in 1978 by Peak District Mines Historical Society, the centrepiece of the museum, then and now, is the 30 feet (9 metres) high Wills Founder water pressure engine based on a design by Richard Trevithick, who also invented the locomotive, and cast at Coalbrookdale in 1819 – one of the oldest and finest engines in the world. Today, however there is very much more at this exciting and much commended museum, with artefacts of mining and miners dating back to the 3000 year old antler tool from Ecton copper mines. Here you can see the tools of lead, spar and copper miners and smelters, learn something of the thousand year old customs made when Derbyshire was still a "wild west" and see something of the miners' lives and the dangers and hardships which beset them. There is even a toy "spar cottage" made in the mid-19th century for a miner's children. The museum has been designed for all ages. It is "hands-on" as much as possible with pumps to wind and simulated tunnels and climbing shafts which delight all young and young at heart.

Over the road from the museum is Temple Mine, where, though there are a few 18th century sections, most of the workings are in a comfortably-sized fluorspar mine reworked in the 1920s and 1950s. Here you can see one of the world's oldest electric locomotives, built about 1934 and used underground in a local mine until a few years ago with the wagons and other tools of the period. Outside the mine is washing and dressing machinery, used to concentrate ores. You can try your hand and learn how to pan for minerals such as gold – though here it is usually only "fools gold" which is found!

Peak District Mining Museum is alongside the A6 road, with parking nearby and is open every day (except Christmas Day) from 11 am to 4 pm, longer at busy times. There is a shop with a large range of unusual gifts and general, local and specialised books. School and other parties are welcome, and there are also special low rates for families and joint visits to museum and mine. Telephone 01629 583834.

A class tries out the reproduction rag-and-chain pump

LATHKILL DALE DERBYSHIRE

• its Mines and Miners •

J. H. Rieuwerts

Landmark Collector's Library

Published by
Landmark Publishing Ltd,
Waterloo House, 12 Compton, Ashbourne, Derbyshire DE6 1DA England
Tel: (01335) 347349 Fax: (01335) 347303
e-mail: landmark@clara.net www.landmarkpublishing.co.uk

ISBN 1 901 522 80 6

© **J.H. Rieuwerts 2000**

The rights of the author of this work
have been asserted by him in accordance with the Copyright,
Design and Patents Act, 1993.

All rights reserved. No part of this publication may be reproduced, stored in a retrieval system or transmitted in any form or by any means, electronic, mechanical, photocopying, recording or otherwise without the prior permission of Landmark Publishing Ltd.

British Library Cataloguing in Publication Data: a catalogue record for this
book is available from the British Library.

Print: MPG Ltd, Bodmin, Cornwall
Editor: Dr T. D. Ford
Design: Mark Titterton

Front cover: Gank Hole Mine, 1884
Back cover: In Mandale Rake
Page three: Mandale Engine House and Shaft

Nomenclature: Throughout this book, Lathkill Dale is shown in this manner. However the sough of the same name was traditionally spelt Lathkilldale Sough and this notation has been kept.
Also the valley was sometimes called Larkhill and this appears where a relevant quotation is used.

CONVERSION TABLE

1 Fathom = 6 Feet
Feet to Metres x 0.305
Yards to Metres x 0.914
Metres to Yards x 1.094

MEER CONVERSION TABLE

High Peak = 32 Yards
Hartington & Granges = 29 Yards
Bakewell = 29 Yards
Ashford = 29 Yards
Youlgreave/Haddon = 28 Yards

DISCLAIMER

The publishers and author accept no responsibility for any loss, injury or inconvenience sustained by anyone entering or approaching old mine workings. They should be regarded as dangerous and avoided.

Contents

Preface	7
Introduction	8
Chapter 1 The Mining Area	12
Chapter 2 The Early History of Mining 1284-1750	25
Chapter 3 The Late Eighteenth Century	35
Chapter 4 The Mandale Company 1797-1839	45
Chapter 5 Power to Drain the Mines: Lathkill Dale Company and Mandale Mine Company	53
Chapter 6 Features Visible Today	79
Chronological Table	91
Glossary of Mining Terms	96
Bibliography	104
References	105
Subscribers' List	109
Index	110

Acknowledgements

I am indebted to His Grace the Duke of Devonshire who has permitted almost unrestricted access to the mining archives at Chatsworth House for some forty years. The archivists at Chatsworth, Mr. P. Day and Mr. Tom Askey are thanked for their considerable help, freely offered at all times. The Barmaster's Collection, although the property of the Duchy of Lancaster, is also held at Chatsworth and to the Duchy and the Barmaster, Mr. William Erskine, I am extremely grateful for permission to consult their collection.

I owe a debt of gratitude to His Grace, the late Duke of Rutland, who allowed access to the lead mining archive at Belvoir Castle.

During research for the first edition of this work, His Grace, the Marquis of Lothian kindly permitted access to the archive at Melbourne Hall. More recently Mr. Howard Usher of Melbourne has been kind enough to provide me with additional extracts relating to late 19th century mining in Lathkill Dale, found as he was preparing a calendar of the muniments held at the Hall.

I wish to thank all the following librarians and archivists: at Bristol Archives; Derby Borough Library; Derbyshire Record Office; Public Record Office, Kew, London; Northumbrian Record Office; Sheffield Archives [formerly the Local History Department, Sheffield City Libraries]; Staffordshire Record Office.

Mr. Michael Cockerton, Steward of the Derbyshire Barmote Courts and Mrs. W. Pearce of Over Haddon have both allowed me to examine documents in their possession and extracts to be published herein; to them both I am very grateful.

As always, my friend Mr. Roger Flindall has been unstinting in providing information from his vast research archive, but specifically the very important detail relating to the large water wheel at Lathkill Dale Mine.

During preparation of the first edition of the book, Dr. Trevor Ford and Dr. Nick Butcher made useful comments about the geology of the Dale.

The first edition of this book contained a considerable amount of information relating to the miners and other personnel associated with mining in Lathkill Dale. This has been retained, virtually unmodified. The information was due largely to research and analysis carried out by the late Janet Wadsworth. Without her help the book would have been much the poorer.

Photographs have been kindly provided by Mr. Paul Deakin; Mr. Harry Parker and Mr. Lindsey Porter.

The plans and sections of existing mining remains in the Dale are due to the good offices of Dr. John Robey and Dr. Steve Thompson.

Lathkill Dale is now managed and controlled by English Nature. Access to the mines is not permitted without previous permission and is not encouraged. Please observe the Countryside Code at all times.

Preface

Nearly forty years ago the author wrote a somewhat brief history of lead mining in the valley of the River Lathkill, near Bakewell, Derbyshire. This appeared in volume 2, part 1 of the Bulletin of the Peak District Historical Society and a small supplement was added a little later. The first edition of the present work followed in 1973.

Several problems, which existed in 1973, have now been resolved satisfactorily, but unfortunately others remain unsolved, and inevitably a certain amount of theorising has been necessary.

I have to admit with some regret that several vital links are still missing. Little information has been forthcoming regarding the activities of the London Lead Company, who operated on a lavish scale in the Dale for a period of twenty years in the second half of the 18th century. The period from 1825 to 1830 is virtually unknown so far as the Lathkill Dale Company is concerned, and there is an annoying gap in the records of the Mandale Mining Company from 1839 to 1852. The relevant books and documents may have been destroyed, but there is always the hope that one day they will turn up in a solicitor's office or in some long forgotten trunk in an attic or cellar.

Introduction

The valley of the River Lathkill is one of the most interesting of the lead mining areas of Derbyshire. Extending from a little east of the village of Monyash, it is a typical, deep, limestone valley, quite rugged and spectacular in stretches, well wooded elsewhere, which eventually becomes less bold below Over Haddon. A little further downstream, after joining the River Bradford at Alport, it eventually meets the River Wye in pleasant meadow land between Picory Corner and Rowsley. The Dale is a favourite beauty spot for miles around and a considerable section of it is now in the hands of English Nature, who own or lease the land within the boundary of the National Nature Reserve.

A hundred and fifty years ago the scene must have been very different. A walk in the Dale then would have revealed not peace and serenity, but a massive Cornish pumping engine with the accompanying hiss of steam as the gigantic beam moved rhythmically up and down, each deliberate stroke vibrating the engine house to its very foundations. Alongside, a water-wheel 35 feet in diameter slowly and ponderously turning, both steam engine and wheel united in the common task of freeing the underground workings from their deadliest enemy – water.

The surface would have been littered with mining implements, spades, shovels, picks, with heaps of the bright gleaming lead ore waiting for the Barmaster to measure it before sale to the smelters. All lead ore was measured before being packed on to the backs of of ponies and taken to the smelting mills. It was measured in a wooden dish of exact volume (calibrated against a Standard Dish). Each dish would normally contain about 65 lb of lead ore and nine dishes made a load, or about a quarter of a ton.

Beneath ground too was a scene

Introduction

of which few people today have but a remote idea. Men, working only by the light afforded by tallow candles, would be drilling the shot holes in the hard limestone prior to blasting. Each hole was hand bored and took anything up to $1^1/_2$ to 2 hours to complete. It was then charged with black gunpowder, the primitive fuse – a length of straw with a little powder in it – was ignited and the rock blasted down ready to be filled and carried away.

A typical level driving under these conditions generally advanced at a rate of about four to five feet per week. In some of the workings men would actually be mining the lead ore. Working in sometimes narrow, but relatively high openings, they laboriously wedged and picked out and sometimes blasted out the precious ore. The measuring and bargain days were certainly 'red letter days' at the mines. In some areas they were known as 'taking days.' A substantial amount of ale would be sent to the mine on these occasions, so that the somewhat laborious task of measuring the ore, prior to sale, was a well lubricated one. The new bargains, struck between the miners and the Mine Agent, for the driving of levels and the mining of ore for the coming weeks, were also carried out on the same day, this latter task being carried out at the local hostelries. The ale consumed on these occasions was provided by the mine proprietors, and several public houses in Youlgreave and Over Haddon were patronised. One can imagine the festivities lasting well into the night, highlighted perhaps with a vivid tale of drama or humour enacted deep beneath the Dale.

Today all is silent. Only an occasional ivy-covered stone wall of massive proportions, the piers of a ruined aqueduct, or the inviting dark entrance to a shaft or level remind one that here was once a major industrial site.

Lead mining was a specialist occupation carried out by insular groups of workers in remote areas of the country so it was inevitable that a whole series of terms and words were evolved that were peculiar to the industry. To avoid giving lengthy explanations whenever they occur in the text the reader will find these special mining terms in a glossary at the end of the book. A far more comprehensive glossary of Derbyshire lead mining terms was published recently (Rieuwerts, 1998).

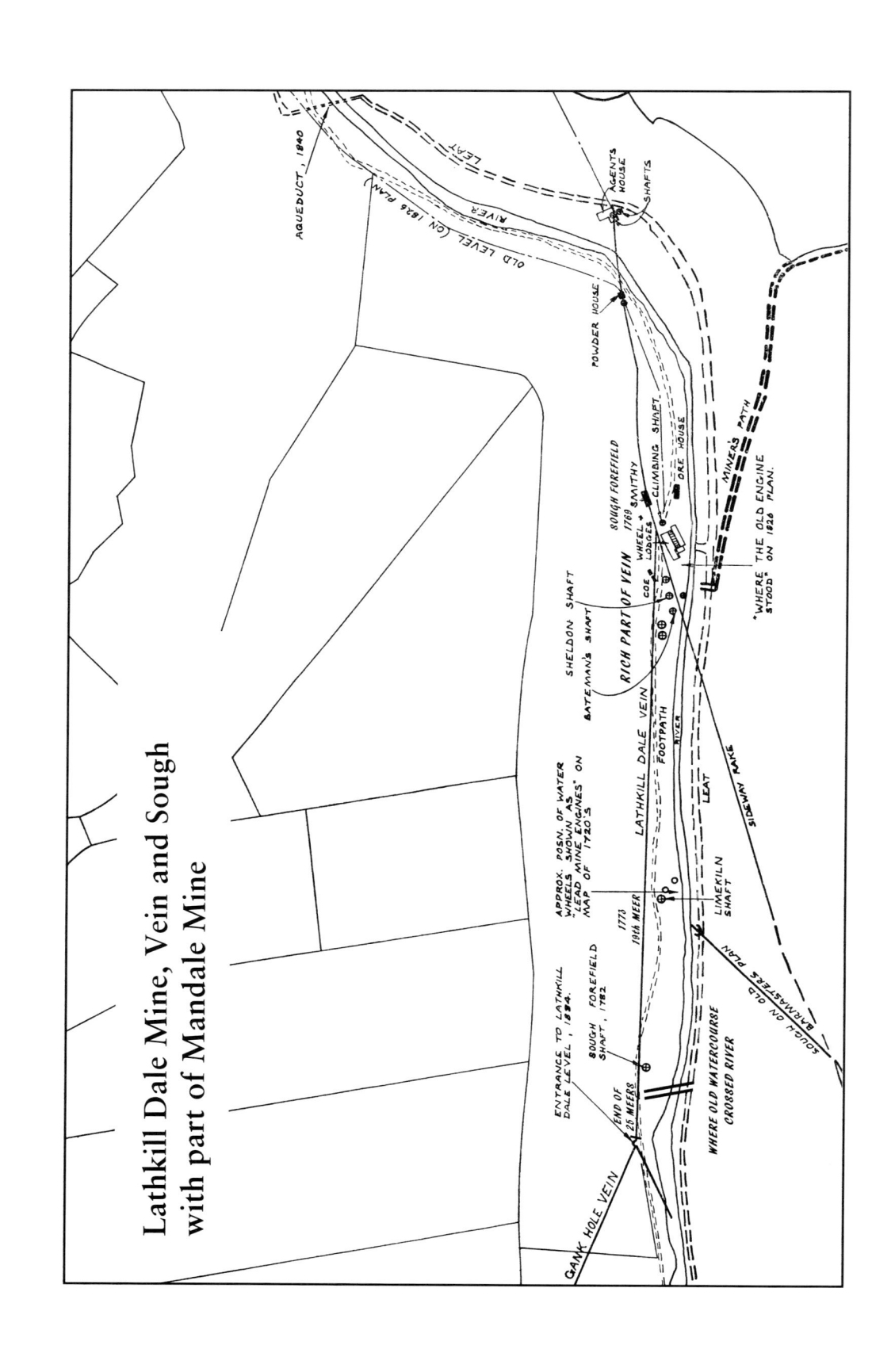

Lathkill Dale Mine, Vein and Sough with part of Mandale Mine

Easterly continuation of the adjacent map

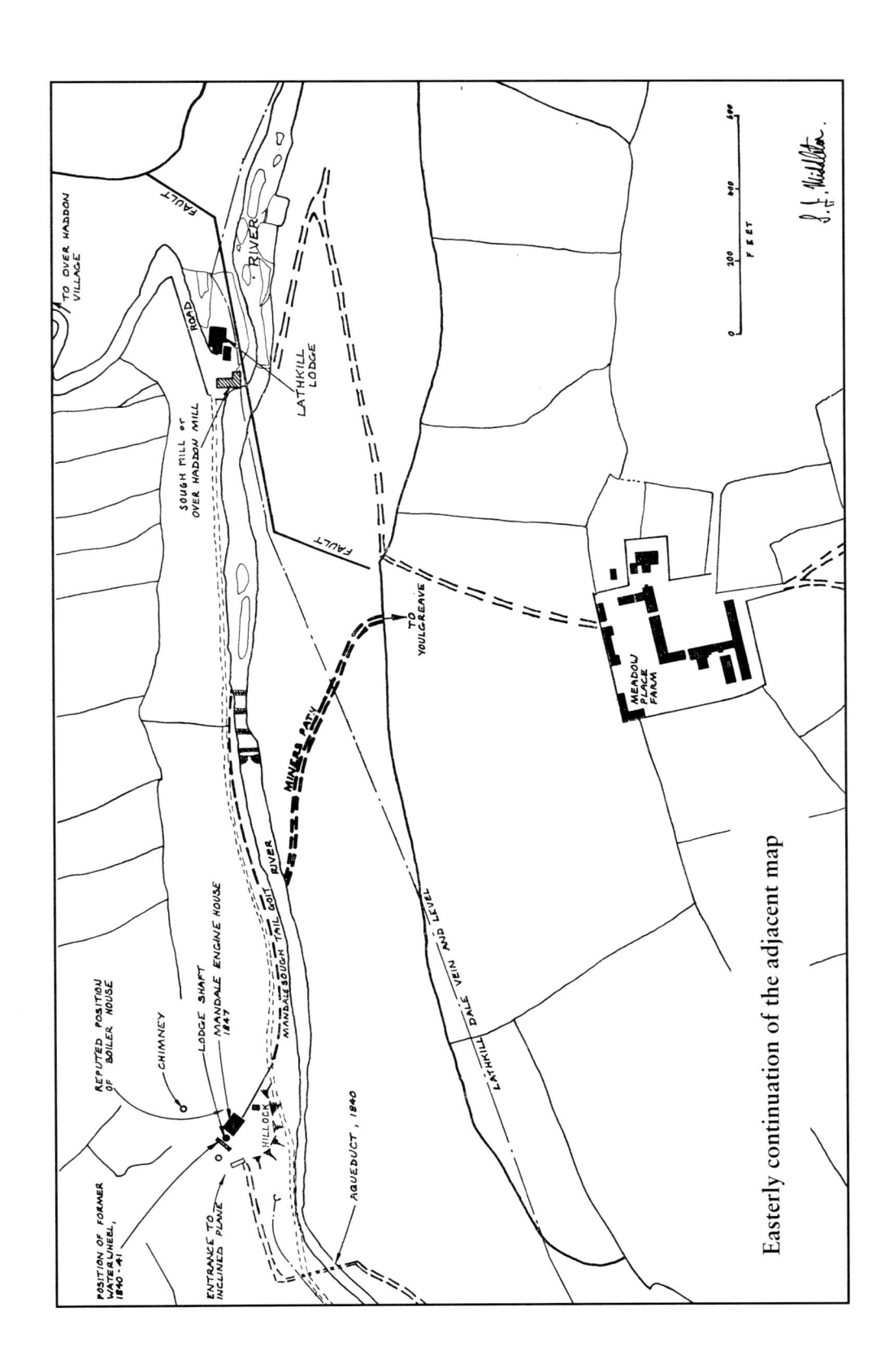

1 The Mining Area

The position of the veins and surface features

Geographically, this book deals with an irregular area of about a mile to a mile and a half in width on both sides of the River Lathkill and extending from Monyash in the west to a little east of the village of Over Haddon, south-west of Bakewell. The area discussed falls within the lead mining liberties of Monyash and Upper (or Over) Haddon in the Queensfield; Ashford Southside Liberty, and Hartington and the Granges Liberty, both of which are now combined into a larger liberty which includes Tideswell and Peak Forest and held by His Grace the Duke of Devonshire; and Bakewell Liberty which is private and belongs to His Grace the Duke of Rutland.

The most important lead veins in the district are the Lathkill (or Lathgill) Dale Vein, Mandale Vein, Pasture Rake and Greensward Rake. Some, for example Ridge Rake, Sideway Vein, Mycross Vein, Gank Hole Vein and Smallpenny Vein have been important for short periods whilst many others, for example Ringinglow Vein, Breeches Vein and Haigs Greave Vein have probably never been worked systematically during their entire history, certainly not below the level of the Lathkill.

Lathkill Dale Vein

This commences at Lathkill Lodge and ranges in a general south-westerly direction through Meadow Place Wood to two large shafts which are now always referred to as 'Bateman's shafts.' This name is misleading because it is not their correct name. A more correct title would be 'New Shaft in the Wood,' at least for the more northerly of the two. The course of the vein through Meadow Place Wood is largely lost in a tangle of dense

1 The Mining Area

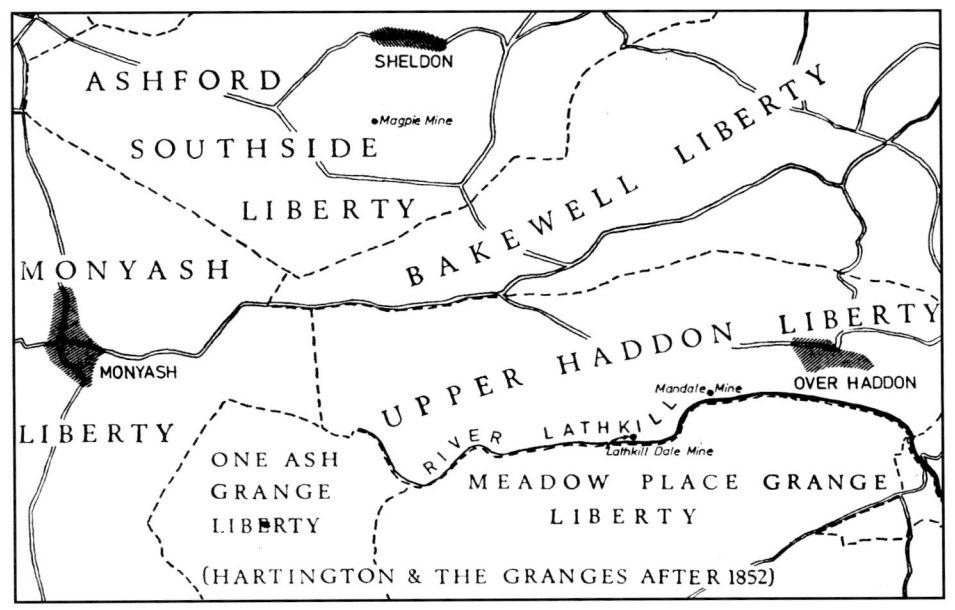

The mining liberties in the Lathkill Dale area (based on Stokes, A.H., 1880-81, *Lead and Lead Mining in Derbyshire*)

undergrowth, although it can be picked out for a short distance in the fields above, as a line of small, grassed hillocks.

The two shafts at Bateman's House are both stone lined at the top. One shaft is somewhat surprisingly situated beneath the remains of the house which was built for the agent of the mines, but it may have had a dual purpose. The significance of this location will be discussed in Chapter 5. There were at least three separate buildings on the site, and they are shown on old photographs. Plans drawn in 1827 do not show any buildings in the area but work on the main building(s) was probably underway by 1830. Expenditure on items such as bricks, plaster and laths and an ash grate indicate that at least one structure was erected in 1835. A sad ruin, it is now completely beyond repair, although stabilisation of the remains is planned shortly. The house was described as being deserted and the garden full of weeds by 1861.

From these shafts the vein crosses beneath the river, and then ranges westwardly approximately parallel to the river and between it and the footpath on the north bank. West of the river the course of Lathkill Dale Vein is marked by a series of run-in shafts and shaft hollows, some partially filled with water. Local information states that some of the shafts along this part of the vein are not filled in, but were merely covered over during the 1930s. The site of the great fifty-two foot diameter water wheel, 290 yards west of Bateman's House, is marked by remains of the breast walling, two pillars of the aqueduct and large shaft hollows. Nothing remains of the 'old Engine' which, though not proven, may have been a very early hydraulic engine. Excavation would, no doubt reveal more. The site is mainly grown over and year by year becomes more obliterated by vegetation. The greater part of the valley floor at one time was covered with old mine hillocks, some of which have only been removed in comparatively recent times.

A long sough was driven along Lathkill Dale Vein. The tail (205662) has run-in, but the course of the level can be traced along the north bank of the river by run-in shafts and denuded spoil mounds, to a shaft in the garden of Lathkill Lodge. From this point surface evidence is scanty, but from information derived from mining plans and from underground exploration, its course is known with some degree of certainty. The most westerly point it definitely reached was the Sough Forefield Shaft (187658).

The Gank Hole Vein breaks out of Lathkill Dale Vein 830 yards west of Bateman's House. Ranging north-westwardly through the wood: little can now be seen, save the limestone walls of the vein and overgrown, backfilled stopes. A level, which served the dual purpose of drainage and haulage, was driven along the sole of the vein in the 1880s, but the entrance is completely obliterated. Remains of a coe can be seen. Photographs taken about 1885-1890 when the mine was in operation, show the layout at the level mouth.

Mandale Rake and Mine

Mandale Rake ranges from Lathkill Dale north-westwardly within Over Haddon Liberty for 60 meers (1920 yards) to the Bakewell Liberty boundary at the Bakewell – Monyash road. The closely parallel Pasture Rake contained 45 meers of ground from the River Lathkill. It ranged north-westwardly to Over Haddon Pasture West Fence where it joined into Mandale Rake.

Principal Veins & Soughs in Lathkill Dale

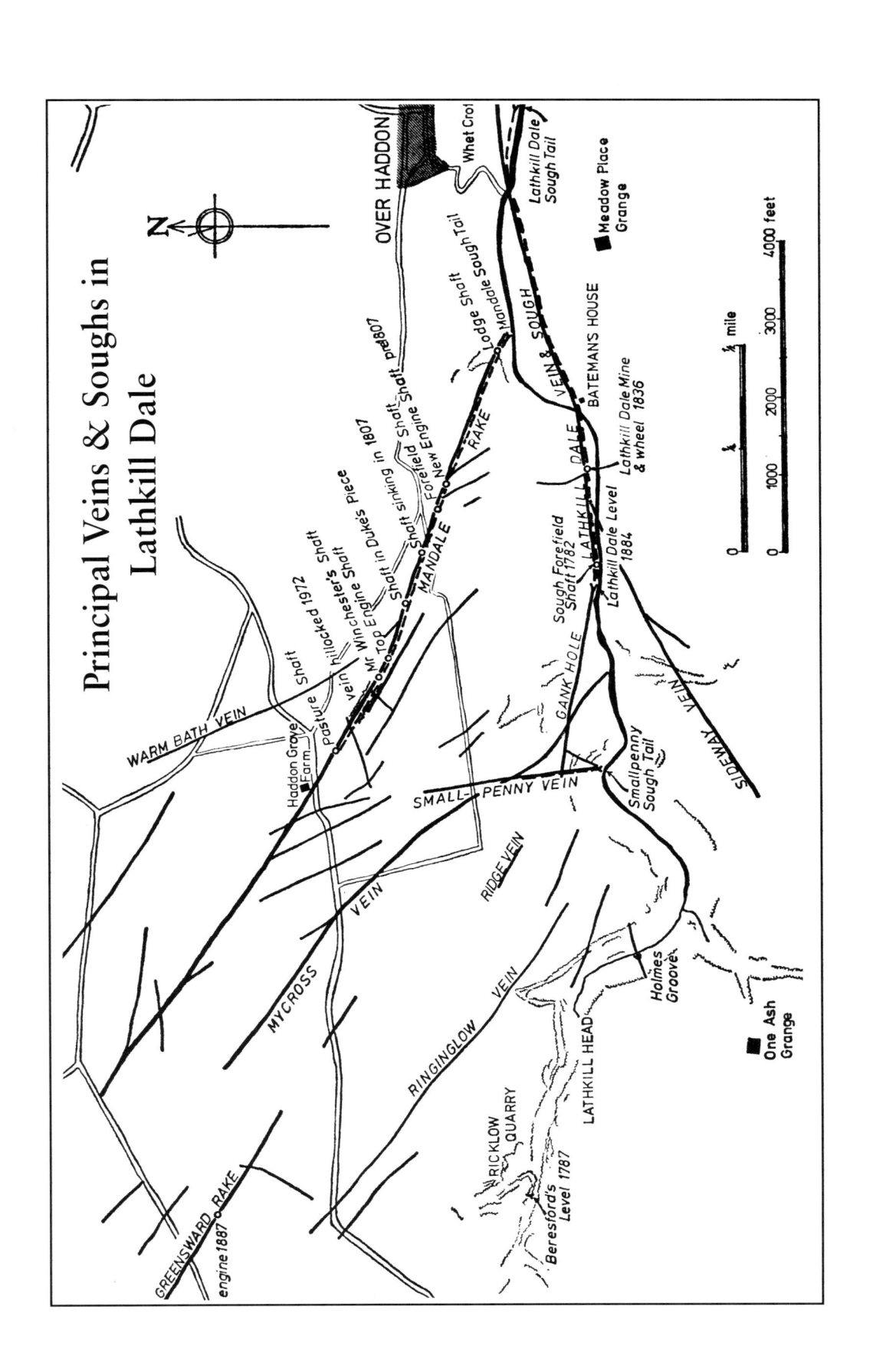

1 The Mining Area

Mandale Mine, as distinct from Mandale Rake, was situated on the rake in the vicinity of the Top Engine Shaft and the eighteenth century Mr. Winchester's Shaft (185666). The old Founder Shaft lay adjacent and it seems highly likely that the mine took its name from Memdale or Mandale, the shallow dry valley that crosses the rake at this point. If so, then the work mentioned in the 1288 Quo Warranto (see Chapter 2) was probably also located in this vicinity. The vein itself ran south-east from here to Lathkill Dale and north-west through Bakewell Liberty into Ashford Southside Liberty, almost to the Hard Rake Vein and the Highlow Mines. At Winchester's Shaft there was a very good walled gin circle or gin-race, 48 feet in diameter. There was also a gin-circle, 30 feet in diameter at the Top Engine Shaft, and a third at the Deep or Forefield Shaft, 42 feet in diameter (190664). This latter was also walled, but like the others has been removed by hillocking (reworking the waste). The shaft is still open, though covered, and was descended in 1965 and 1973 (69).

The Mandale Mine Reckoning House, referred to as the 'New Reckoning House' in a letter dated 1826, (1) appears to have been what is now a barn at 184666. A good deal of hillocking has taken place hereabouts, but it was interesting in revealing that all the top of the vein was worked opencast, and subsequently backfilled with refuse vein-stuff from the dressing floors. The few remaining surface features along Mandale Rake display the usual grassed hillocks.

At the south-eastern extremity of the Mandale Rake and Pasture Rake Vein in Lathkill Dale, is the site of the nineteenth century Mandale Mine. Here more interesting remains are to be found. The Mandale Sough is open and together with the main workings accesible via the Inclined Plane and the 'Aqueduct Level' give a total of over 600 yards of levels and stopes which can be explored. (This figure does not include numerous climbs into old roof workings). The sough is also accessible from the Deep or Forefield Shaft. A connection has been established between here and the workings in from Lathkill Dale and the sough level is also accessible for some distance northwest of the Forefield Shaft (69).

The Cornish Engine House is in poor condition and in reality only the 'bob' wall exists in any detail. Excavation of the engine house may be undertaken in the future. Behind the engine house is the impressive, horseshoe shaped

Above: Oil painting of the Mandale Mine aqueduct in 1892. The wooden troughs or launders were presumably sold when the mine was abandoned. **Below:** Nineteenth Century photograph of the piers by Richard Keene. He appears in the photograph.

Lodge Shaft from which the Cornish engine pumped water into the sough, as also did the earlier water wheel. Some portions of the retaining walls associated with the wheel installation still remain. The boiler house has been completely obliterated, but the ruined flue and base of the former chimney can still be seen on the nearby hillside.

Robinstye Mine

The eastern end of Lathkill Dale Vein is also the location of the problematical Robinstye Flat Work and Rake. Only four references are known, three from one source. The first is from a Barmaster's Book of entries (2):

> '15.2.1809 ten meers of ground in old vein called Robinstye Vein, all ranging eastwardly from an old hillock in a piece of land called the Whet Croft.'

Whet or Whot Croft is a field immediately east of St. Anne's Church, rising steeply from Lathkill Dale towards Over Haddon village. The field contains two shafts, both now covered with concrete slabs. These shafts could well be on Robinstye Vein. The other three references are all contained in Farey's survey of 1811-1817 (3). From his list of mines:

> 'Robinstye Flat Work and Rake, in Meadow Place Grange and Over Haddon, in 1st Lime and Toadstone, the Great Bakewell Fault crosses it.'

Later in his list of springs of water he lists a large one:

> '... in Robinstye Mine, where crossed by the Great Bakewell Fault.'

Finally he comments that this Great Bakewell Fault

> '... proceeds in the 1st Lime west of Youlegreave, Conksbury and Meadow Place Farm and crosses Robinstye Mine and the Lathkill south west of Over Haddon'

These four references, with the addition of an old mining map which shows the range of the fault across Lathkill Dale at Sough Mill, seem to make it reasonably certain that Robinstye Vein ranges south-west to north-east approximately parallel to Lathkill Dale Vein. Alternatively, it may even be the north-easterly continuation of Lathkill Dale Vein itself on the north side of the river.

Other Mines

Sideway Vein can be traced in a south-westwardly direction through

Meadow Place Wood, by a series of run-in shafts, small open shafts and narrow open stopes in the vein. At 186656 is a beautiful stone-lined buddle. A large, conspicuous level on the south bank of the river at 188658 is shown on one plan as being a sough driven into the Sideway Vein. The level is more likely to have been driven as a drawing gate, but water may have drained along it.

Small, partially run-in levels are visible on Mycross Vein, Ridge Rake, Gank Hole Vein and Smallpenny Vein. At the latter, the run-in tail of the Smallpenny Sough, mentioned briefly in 1814, is still vaguely visible by the side of the ruined coe. The late Mr. Charles Millington could remember seeing the old sieve used by the miners for riddling and sizing their ore, lying in the coe 100 years ago. He believed the vein received its name because the ore after dressing was 'only th' size of a smaw penny'.

Geology and Mineralisation: A Brief Note

The valley sides of Lathkill Dale, from east of Sough Mill westwardly to Lathkill Head Cave, are seen to consist of an almost continous section exposing the Upper and Lower Lathkill Limestone (4).

These limestones are now referred to by the British Geological Survey as the Monsal Dale Formation. Overlying the Lathkill Limestones and outcropping west of Lathkill Head Cave, and elsewhere, are the Eyam Limestones perhaps best exposed in the old Ricklow Quarry. However, these latter beds are not really relevant to the mines or mineralisation in that portion of the Dale under discussion.

Igneous Horizons

A little east of Over Haddon Mill or Sough Mill, a layer of basalt lava or 'toadstone' is seen on the north side of the valley, and the small shaft mounds along the line of Lathkilldale Sough between here and the sough tail are also composed of the same material. The basalt represents two separate lavas within the Upper Lathkill Limestones (5). The lavas are at this point brought up against the limestone by a fault trending southwest-northeast across the valley at Sough Mill (4). The fault is shown on an old mining map (6) and is the 'Great Bakewell Fault' proposed by John Farey in 1811. The following extract from Farey is worth quoting, though the fault is of relatively little significance according to modern mapping:

1 The Mining Area

'proceeds in the first lime west of Youlegreave, Conksbury and Meadow Place Farm and crosses Robinstye Mine and the Lathkill southwest of Over Haddon, where the first toad-stone, the second lime and again the first toadstone abut east on this Fault for short distances, with the first lime on the west side'.

Farey recorded that ore was worked in the toadstone in Dale Mine, Over Haddon, and he was very probably referring to the eastern range of Lathkill Dale Vein.

No toadstone was met with in the workings of Mandale Mine, but a reckoning book speaks of a level being driven 'on the 60 fathom Clay.' This suggests a clay wayboard, either very near to the base of the Lower Lathkill Limestone, or possibly a little lower.

William Hopkins, author of *The stratification of the limestone district of Derbyshire* published in 1834, mentions this clay in a letter

The former Carter's Corn Mill in the late 19th Century

addressed to William Wyatt, the well known lead mine agent and mine owner:

> 'Oct. 13th, 1838... in the mines about Sheldon no toadstone was found, except the sixty fathom clay could be considered such.'

He goes on to comment in some detail regarding the position of the so called 'First Toadstone' and 'Second Toadstone' of Farey, in the region of Flagg, Monyash and Sheldon (7). These are now thought to be beneath the Lathkill Limestones.

The Mandale Rake

This is a large fault fissure showing several phases of mainly horizontal movement. The veinstuff is composed of thin strings of galena in a gangue of predominantly pink and white barite, with some calcite, the whole being associated generally, but not always, with a great deal of brecciated limestone fragments in a matrix of brown and orange-brown limonitic clay. Parts of the vein are vertical but some sections lean or hade. The vein varies in width from a thin 'stringer,' half an inch wide, to large 'bellies' or 'boxes,' 16 feet or more in width. Mandale Mine is variously described in old documents as both a rake and a pipe. Examination of the workings reveals that some of the larger stopes do not always have easily definable vein walls. The limestone is often very shattered and mixed with limonitic clay, the whole crumbling away quite easily to the touch. A report made to the proprietors of Mandale Mine in 1842, commented on 'the extreme rotteness of the Vein.'

However, in several sections it is possible to examine workings immediately above or alongside one of these boxes and they are found to be the usual, small, typical old lead mine stopes, quite narrow but with well defined hard limestone walls. The vein is seen to be a barite 'stringer,' perhaps only a few inches in width, with a little galena (lead ore). Thus the nature of the vein can vary in character, both vertically and horizontally over a short distance, changing from a large pipe cavity to a thin scrin.

From the bellies or boxes the great ore strikes of 1820 and 1823 were obtained. They were found where, according to an old section of the mine, (8) the vein passed through a bed of ' Blackstone,' the miners' name for a dark-coloured, bituminous limestone. The term 'Blackstone' must be interpreted with caution; for example in the Matlock area it often refers to toadstone. At Mandale Mine, it almost certainly is the equivalent of the Lower Lathkill

The remains of an aqueduct dating from 1840 are a well known feature of the dale.

Limestone, with perhaps the basal beds of the Upper Lathkill Limestone. The Lower Lathkill Limestone is largely equivalent to the Monsal Dale 'Dark Beds'. Exposures underground show it to be thinly bedded, usually dark coloured with prominent chert horizons. The bed underlying the 'Blackstone' is shown on the old section as the 'Greystone,' and probably corresponds to the lowest beds within the Monsal Dale Limestones but it does not outcrop in Lathkill Dale.

Lathkill Dale Vein

The exposures along Lathkill Dale Vein are extremely limited, so that detailed inspection is impossible. The vein at the bottom of the shaft at Bateman's House is about 2 feet wide, composed mainly of cream-coloured calcite with broken limestone. The walls, which are well defined, display horizontal slick-ensiding and are in very thinly bedded, dark bituminous limestone, no doubt the 'Blackstone' of Mandale Mine. The vein hades to the north at an angle of about 20°-30°.

Other Minerals

At the western end of Lathkill Dale Vein, a vein known as the Gank Hole breaks out in a north-westerly direction. This vein was worked intermitently as a source of limonite (iron hydroxide), used mainly as a pigment. Samples were sent for analysis with a view to the mineral being used as an iron ore. Dr. A. J. Bernays, to whom the samples were sent, commented in a letter, 20th January, 1855:

> 'The iron ores you sent me would be very good if they were uniform . . . I have selected two fair specimens. The one I have called ochrous on account of its porosity and colour. I find . . .54.07% metallic iron. The other specimen, which is very dense is better still . . .58.99% metallic iron' (9).

Hematite ore is said to have been found in the Stone Pit Rake Mine. Limonite can also be picked up along the Smallpenny Vein. Goethite (iron hydroxide) is present in Mandale Mine and elsewhere.

South of Horse Lane, smithsonite (zinc carbonate, commonly known as calamine in Derbyshire) could be found in small quantities in the north-western end of the Mandale Rake, before hillocking took place.

The mineral rosasite, an uncommon carbonate of copper and zinc, has been identified in a small level along Gank Hole Vein (10). Identification has not been made in any other mine in the Dale.

In 1854 gold fever hit Lathkill Dale with reports that the precious metal had been found in the lava at a newly opened lead mine at Cow Close, east of the village of Over Haddon, in a field between the village and Conksbury Bridge. Suffice to say the bubble burst in 1856 and it is extremely unlikely that gold in any but the most minute amount was found. It was supposed to have been contained in iron pyrites in the toadstone, and as this latter mineral is some what sparingly distributed in the toadstone or lava it becomes obvious that little, if any, gold exists in the area (11).

Fluorspar, whilst not a major gangue mineral in Lathkill Dale, has been recovered from the old hillocks of the Mandale Rake, between the Top Engine Shaft and Haddon Grove Farm.

The ore and gangue appears to have been more concentrated in the more massive limestones of the Upper Lathkill Limestone series, with poorer values in the thinner bedded limestones. This explains why fluorspar was much richer near to the surface of Mandale Rake for instance.

Above: Old hillocks along Mandale Rake, looking north westwards. The 13th Century workings at Mandale were probably very near these hillocks. **Below:** Miners' stile giving access to Mandale Rake. It is not a public footpath.

2 The Early History of Mining 1284-1750

Ancient Mining

Traditionally, one of the county's earliest workings was the Mandale Mine and the mine is specifically mentioned in the well known 'Quo Warranto' (12), an inquisition held in 1288 at Ashbourne to enquire into the Derbyshire miner's right to dig for lead ore. Apparently all purchasers of lead ore had been prevented from buying at the Mandale Mine for a period of four years previous to the 'Quo Warranto'. There has long been a story amongst local folk, particularly old miners, that the mine was so rich at this period that the market was flooded with cheap lead ore thereby forcing smaller mines out of existence.

Documentary evidence to this effect is not too convincing, although in 1287 a Jury had met at Over Haddon. One cannot be sure whether there is any connection between the trouble over the ore buyers at Mandale, such buyers incidentally generally paid a duty known as cope as a pre-emption on all lead ore purchased for smelting, and the enquiries of the Jury in 1287. What is proved of course is that mining was taking place in and around Lathkill Dale at least by the end of the thirteenth century. One of the questions into which the Jury had to inquire was why William de Hamilton, owner of a third of the mineral royalty in Taddington, Priestcliffe and Over Haddon, had this detained from him. Also, why his bailiffs, who had wished to buy cope ore on his behalf in Over Haddon, were:

> 'disturbed by Nicholas de Cromford and Simon de Cromford so that they could not buy it during the time aforesaid.' (13).

Four years after the Quo Warranto, in 1292, there is a further reference to lead mining (14). The Abbey of Leicester held considerable property at what is now called Meadow Place Grange. At that time it was known as 'Medoplek' and they owned mines which were valued at 2s. per annum. The mill was considered to be worth 10s. per annum.

Mining in the 16th Century

Lead was probably mined on this south side of the river for some considerable length of time, because during the early part of the sixteenth century there was a dispute between the landowners of Over Haddon and the Abbey over the boundary between their lands. Both parties claimed the right to Common Land on the south side of the river Lathkill. Amongst the witnesses who gave evidence on behalf of the Abbot was:

> John Weyne of Gratton who gatt oor xxxiv yeares ago on both sides ye water of Lathkyll and meted (measured) on ye south side by ye Abbott Dyshe and on ye north side by ye Kings Dyshe (15).

This reference to the Abbot's dish is most interesting. the date of the above reference is 1527-28, so Weyne did his mining around 1495.

No further details of the Abbots Dish have come to light, by 1649 the dish for measuring ore in Meadow Place Grange was to be identical with the one used in Ashford Liberty.

The portion of Mandale Rake between Winchester's Shaft and Haddon Grove, part of which was hillocked during the 1960s, is known to have been worked opencast to a depth of at least 20 to 25 feet for the entire width of the vein and subsequently backfilled mainly with washed refuse. This mining cannot be dated, but is obviously of great antiquity. In 1585, Over Haddon Field Rake was described as containing the best ore in the Peak.

Into the 17th Century

By 1615, some of the workings at 'Mem Dale' (presumably Mandale) were 300 feet deep (17) and mining was taking place in Over Haddon Pasture in 1632. Mining in and around the Dale was presumably of some importance because about 1635, shortly after the estate had passed into new ownership, an enquiry was made whether it would be profitable to erect a lead smelting mill on the river, or a walke-mill (fulling mill). The agent reported that there was a scarcity of wood for a smelting mill, so therefore a walke-mill

2 The Early History of Mining 1284 – 1750

would be a better proposition. However there are no further references to either, so presumably nothing came of the idea (16).

Lead ore from Mandale Rake was worth 16/- per load in 1640, while a trespass into two meers in Mandale Rake and which were being worked 90 – 120 feet in depth, resulted in 700 loads of ore being mined illicitly; it was valued at £700 in 1677.

A petition was presented by two miners, George Bohme and George Glossop, to the Earl of Devonshire in 1649, relative to their mining rights in Meadow Place Grange and subsequent to this, eleven articles were drawn up. Some of the articles were similar to those in general usage throughout the Peak mining fields, but some deserve special mention. As noted above the method of measurement of ore had to be based on a dish identical with that used in Ashford Liberty:

> 'the oare to be measured by the Barremaster by a dish or gage to be made by the dishe now used by the Barremaster within the Mannor of Ashforde.'

They were to have 29 yards to each meer, which although not stated in these articles is again identical with Ashford Liberty.

Another article, this one unique to Meadow Place Grange, was that when a new vein was found, the miner:

> 'shall have one half of the first three mears for the finding ... the other half of the three mears to be for the use of Earle, the Barremaster paying equall charge with the finder or finders for the workmanship of same.'

If a mine was unworked, the Barmaster could, after giving notice to the owners, dispossess the same after only ten days instead of the usual three weeks required elsewhere. Apparently Barmote Courts were also held within Meadow Place Grange for the settling of any disputes and the usual mineral matters. This document is endorsed 'My Lord of Devon's advice for the Myners at Meadow Place' (18).

In 1679 Cornelius Dale of Monyash (or Flagg) was the Barmaster of Over Haddon. (19). On the 4th October 1694, the Great Barmoot Court met at Upper Haddon (20) and added a 46th article to the High Peak Laws and Customs. This article became the fiftieth in Hardy's *Miners Guide* of 1748.

2 The Early History of Mining 1284 – 1750

The Earliest References to Drainage

The first drainage level in the district to be mentioned by name is Over Haddon Suff in 1730, but before this, and certainly before 1727, two water wheels had been erected on Lathkill Dale Vein, at a point 1650 feet eastward of the site of the nineteenth century Carter's Mill. These two wheels are shown on an old estate map dated approximately 1720 – 1727 (22). The site is merely referred to on this map as 'Lead Mine Engines' and the two wheels are shown side by side on the north bank of the river. Behind the wheels are two mine buildings. No shafts are marked, neither is the position of Lathkill Dale Vein.

The Dale has been greatly altered during the space of the intervening 275 years and examination of the site has not been particularly revealing. Between the river and the footpath is a flattish area six to eight feet above river level, with definite traces of a gin-circle, a large depression on its western side, presumably indicating the position of the gin shaft, and a little further away a smaller climbing shaft. The large shaft is probably

Detail from a plan of 1720-7, showing two water wheels on Lathkill Dale Vein (see text above).

2 The Early History of Mining 1284 – 1750

Limekiln Shaft, named on a mine map of the 1830s. The site is much overgrown and no excavation has been carried out. Immediately east of this site is a lower flat area, almost at river level. No further documentary evidence has come to light regarding these water wheels, or anything of the actual mining itself.

Over Haddon Sough

'Over Haddon Suff' is recorded in an ore account book during the early months of 1730 (23). Unfortunately this old level cannot now be located, although there are several suspicious openings in the Dale which could be the entrances of old soughs. An unnamed sough is shown on a mine map of 1826 (24) and again in greater detail on a map of about 1836-40 (6). This sough drained the upper workings of the Lathkill Dale Vein, westwardly from what later became the site of the powder house, and extended a little over 180 yards west along the vein. Downstream it followed the curve of the river for approximately 360 yards and had its outlet a little east from the Mandale Aqueduct. The course of the level north east from the powder house is marked by a series of deep, water-filled depressions, no doubt former shafts. The tail is not visible. This sough is only called 'Old Level' on both plans.

The richest workings during this period appear to have been Over Haddon Pasture Grove, where during 1727, 42 loads of ore were measured. Small quantities of ore were measured at Upper Haddon Dale, and other mines recorded include Over Haddon Pinfold, Ricklowdale and Ringinglow (23). Amongst the Devonshire Collection at Chatsworth House is an interesting document naming working miners in several liberties including Upper (or Over) Haddon in 1727 and 1728, and those who served on the Grand Jury of the Barmote Court (25). One of these men, George Redfearn, was working at another early drainage level, Pasture Sough, and in 1731 obtained 5 loads and 2 dishes of lead ore there. He served on the Grand Jury in 1723. There is no documentary evidence to confirm its position, it could be another name for Over Haddon Suff, but this seems unlikely. The level known to modern explorers as the 'Aqueduct Level,' driven along the sole of a vein, possibly part of Mandale Vein adjacent to Pasture Rake, is distinctly sough-like in certain sections. Parts of the level are roofed with stone slabs and immediately inside the entrance are two very large diameter

(cont'd on page 32)

2 The Early History of Mining 1284 – 1750

An old trial level just west of Sough Mill. This very short level may be of considerable antiquity.

2 The Early History of Mining 1284 – 1750

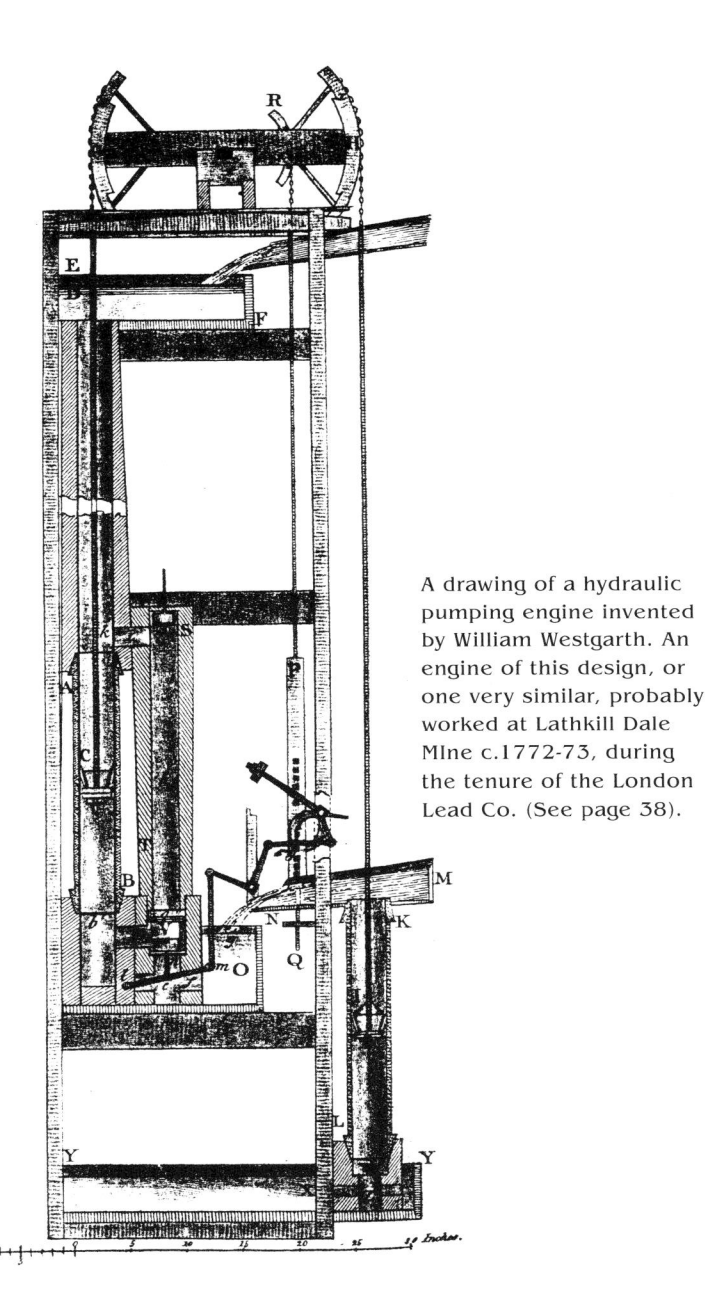

A drawing of a hydraulic pumping engine invented by William Westgarth. An engine of this design, or one very similar, probably worked at Lathkill Dale Mine c.1772-73, during the tenure of the London Lead Co. (See page 38).

shot holes, unlike the usual smaller type seen in most mines. Pasture Sough Vein is mentioned again in the 1740s.

Lathkilldale Sough

A reference which appears to relate to Lathkilldale Sough is to be found in the *Philosophical Transactions* for the year 1744-45 (26). A letter dated May 1744 gives details of a human skeleton discovered:

> 'at Lathkill Dale . . . as the Workmen were driving a Sough or Drain to a Lead Mine, about 9 yards deep and 40 fathom from the beginning of the Sough . . . The place where these things were found is on every side surrounded with a rocky petrified substance or Terra Lapidea, by the miners called Tuft, so hard (as they say) to strike fire against their tools. The substance lay above the Bones . . . a yard and a half thick, and on either side . . . there being a foot of soft coarse clay or Marl interspersed thick with little petrified Balls of the same kind of substance as the Tuft.'

These details were sent originally from a Mr. Platt. The soft, coarse clay is no doubt decomposed lava, and the 'Tuft' is either tufa or possibly basalt lava, particularly plentiful around the tail of Lathkilldale Sough. The comment on the hardness of the tuft is interesting. Calcareous tufa would not strike sparks, but lava might. Lathkilldale Sough is also mentioned in a trial for debt in the Barmote Court held at Upper Haddon in April 1744, when the defendant was John Gould (27). The action obviously relates to work done within Upper Haddon Liberty, and must therefore have taken place between the tail of the sough and where it enters Meadow Place Grange on the south side of the river in Hartington and the Granges Liberty. This, coupled with the reference from the *Philosophical Transactions,* suggests that Lathkilldale Sough may have been started in 1743.

The Bristol Company and some involved Barmote court cases

A most interesting period is now reached, characterised by a large number of trials in the Barmote Court (27, 28). These trials began in 1744 and lasted until 1749. The com- plaints were mainly concerned with non-payment of shares and wages, and for materials delivered to

2 The Early History of Mining 1284 – 1750

the Ridge Rake Mine and Sough, Pasture Rake and Pasture Sough Rake.

The fascinating aspect of all this is the fact many of the shareholders were Bristol businessmen, some of them Quakers. Whether there is any connection between this venture and the London Lead Company's Derbyshire activities is not

The tail or entrance to Lathkill Dale Sough.

known at the moment. One document specifically refers to them as 'Mr. Trevaskes and the Bristoll Co.' Their agent was John Roberts, probably of the Roberts family of Taddington several of whom were Barmasters or Deputy Barmasters. Amongst the complainants were Hugh Sheldon, Charles Potter who delivered wood to the mines, and John Rowland who delivered an 'Engen' worth £7 7s. to Ridge Rake Mine. William Wildgoose and Richard Glossop were both owed wages at Pasture Rake, while Thomas Edwards, James Shemwell and Charles Gould (or Gold) were owed for work done at Ridge Rake.

The shareholders included John Trevaskes, Mathew Wayne (or Waine), John Noble, William Garlick, Robert Denham, John Simons, Marcellus Osbourne, Samuel Lankford, Richard Foxlow, Humphrey Simons, John Gould and Charles Gould. From the above lists of names at least the first five were Bristol merchants, or had connections there (29). The ore accounts for the period do not indicate any significant returns from any of the mines, so one could speculate whether the large number of shareholders is an indication that the mining was largely exploratory. Lathkill Dale in terms of mid-eighteenth century communication was fairly remote from Bristol, but Bristol merchants appear to have been concerned in similar ventures all over the country during the eighteenth century. Perhaps in this instance they were financing the driving of exploratory drainage levels from the river, one along Ridge Rake, one along Pasture Rake, and perhaps also Lathkilldale Sough.

A bill brought into the Barmote Court by John Roberts as agent for Mr. Trevaskes and the Bristol Company against two other shareholders, Marcellus Osborne and John Gould, for the sum of £30 3s. 3d. was resolved by the Jury awarding Roberts and the rest of the Bristol partners the shares which had belonged to Osborne and Gould. Matters were further complicated when, in 1745, a cross bill was brought into the Barmote, but an indication that the mining was a failure may be gathered from the statement that: 'they have expeded considerable sums of money thereat without reaping any benefit or advantage therefrom and are now about to desert the same as a barren and unprofitable undertaking' (30). The initial link between the Bristol undertaking and Derbyshire lead mining, whether through the London Lead Company or not is still unsolved.

3 The Late Eighteenth Century

The Period From 1750 to 1765

The failure of the Bristol adventurers to locate profitable ore at Ridge Rake, Pasture Rake and perhaps along Lathkill Dale Vein, meant a return to the more typical Derbyshire mining – the small venture perhaps worked only on a part time basis.

One name does figure rather prominently however, that of Richard Glossop, a miner and farmer of Over Haddon. His name first appears in 1737 when he raised three dishes of smitham (finely crushed lead ore) at Over Haddon Dale. He worked at Pasture Rake in 1745, when he was owed for 'work done' by the Bristol Company and by 1750 was a partner there. On the 2nd July, 1754 he and William Goose, (probably Wildgoose) another Over Haddon miner, were allowed 6 meers of ground on Sellers Sough Old Vein in Nether Haddon Liberty (31). Previous to this date, in 1751, he began working Sideway Vein in Meadow Place, which by this time had been incorporated into Hartington and the Granges Liberty. He continued to work this vein until 1761, freeing a new title in 1758 which he called New Sideway. During this ten year period he mined 131 loads of ore, the highest yield being 9 loads and 2 dishes in 1760 (32). His mining therefore was limited, but he obviously worked more than one mine at a time, which perhaps proved to be a profitable venture parallel to his farming activities. He served on the Jury of the Barmote Court for Youlgreave Liberty in 1773 and 1774, and died in 1783 (33).

One Elias Glossop, described as a miner of Upper Haddon, died in September 1739 leaving in his will 1s. to his daughter Anne, the wife of William Wildgoose, a miner of Upper Haddon. William Wildgoose was involved at Pasture

3 The Late Eighteenth Century

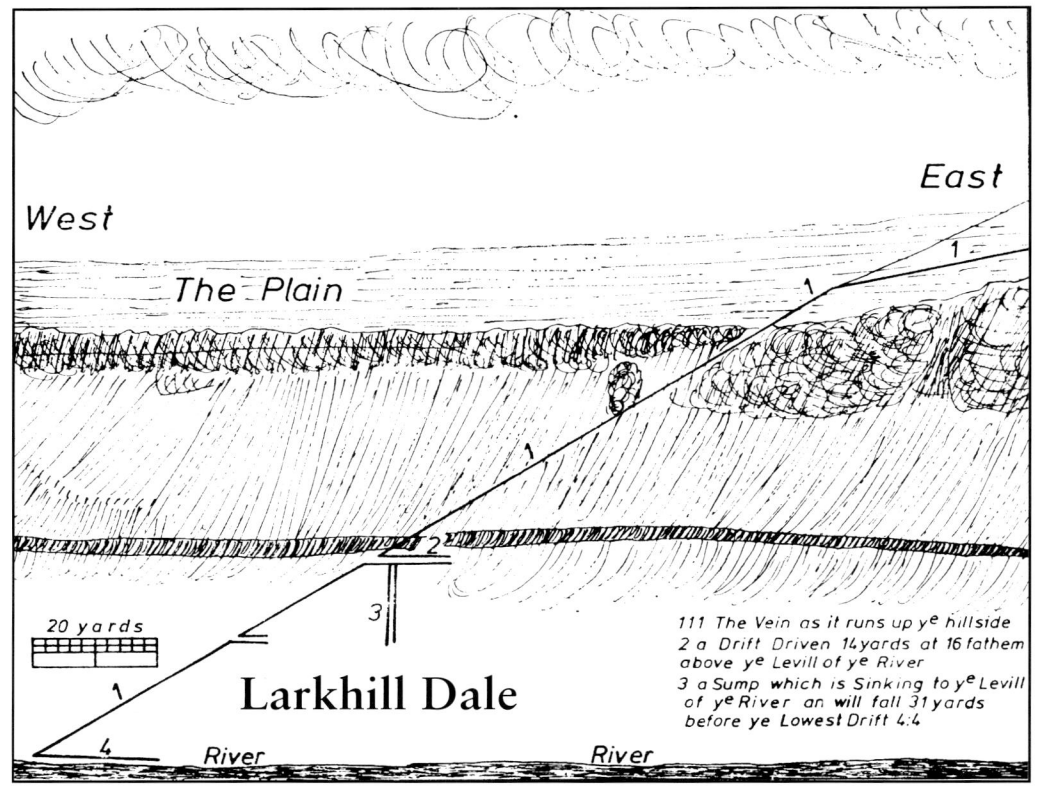

George Heywood's section of an unlocated mine in Lathkill Dale, 1754-5. (Based on Bag 512, Sheffield Archives).

Rake, along with Richard Glossop, who were both owed wages by the Bristol Company. During this same period several other small mines were in production, returning very small quantities of ore. These included Pasture Rake, Ridge Rake, Call-inglowe, Lathkill Dale, Over Haddon Pasture, William Taylor's Over Haddon, Overcloud Vein, Breeches Vein and Meadow Place. The total amount of ore mined by all these mines over an eleven year period was only 127 loads 4 dishes, amounting to little more than 30 tons.

Lathkill Dale Vein is recorded briefly in 1754 and 1755 when George Heyward measured 10 loads, 3 dishes of ore.

George Goodwin the younger, of Monyash, staked out 31 meers of ground in Mandale Rake in Bakewell Liberty on February 12th, 1765, and on the same day Ralph Wheeldon and William Bonsall, two Monyash miners, claimed ground on Milecross (or Mycross) Vein in the same Liberty (31). Although these mines are nearly a mile from Lathkill Dale, they eventually became part of the 'Lathkill Dale Title', and are linked in other ways.

Heyward's cash book for these years contains a section of a mine somewhere in Lathkill Dale, but it cannot be fitted to anything to be seen today (34). An un-named vein is shown ranging approximately north-eastwardly up the hillside and a level driven from the riverside, along the line of the vein for about 20 yards, with a drift 14 yards long at 16 fathoms above the level of the river. A shaft was being sunk to meet the lower level. A thin, but distinct bed of rock is shown immediately above the higher of the two levels, with presumably what is meant to be a line of 'tors' at the valley top. Above this is the 'The Plain'.

Lathkill Dale Sough was still in work, and by 1750 had been driven into Meadow Place Grange Liberty. Four dishes of ore were measured in the sough title in that year (32). There is a local tradition that a sough was driven westwardly beneath the River Lathkill by Cornish miners. There is no direct documentary evidence to support this statement, except that the surname Trevaskes has an obvious Cornish connotation, although his name is not directly involved with the Lathkilldale Sough in any of the Barmote Court trials.

Lathkill Dale Vein was next freed in 1763 by William Wildgoose, but he only measured once, a trifling quantity of ore.

The Era of the London Lead Company 1761-77

The early history of this Company's activities in Derbyshire, particularly in the Winster area, is well known and has been recorded by the late Dr. A. Raistrick (35) and others, including this author (confidential report to Peak Park Joint Planning Board). They subsequently owned mines elsewhere in the county, and during the 1760s began to take an interest in Lathkill Dale and adjacent areas.

The title to Smallpenny Vein was acquired in 1761, but only one measurement of 6 dishes of ore was made. This, so far as is known, was the first title to be taken in the district. There follows a brief silence until the 27th of June 1764 when the title to Lathkill Dale Vein was taken (32).

Analysis of the ore accounts reveals that from June 1764 until June 1769, the maximum output for any year was 266 loads measured in 1766. Then, from June 1769 until exactly one year later no ore at all was measured, immediately after which 874 loads were raised in six months. The next three years were also good, 844 loads for example were measured in 1772.

A mine plan dated 1826 shows an 'old watercourse' commencing close to the New Mill (Carter's Mill), crossing the river to the south bank and leading to what is termed 'place where the old engine stood' at the position of the nineteenth century Lathkill Dale Mine (24). The type of engine is not specified, but there is some evidence to suggest that it might have been a water-pressure engine of the type invented by William Westgarth. Several of Westgarth's engines were installed at mines in the Northern Pennines from the late 1760s. This is a very exciting proposition, because no other Westgarth engine is known in Derbyshire. It is difficult to envisage that this engine, whatever type it was, could have been erected by anyone but the London Lead Company. Whilst it is appreciated that two water wheels were working during the 1720s and that the Bristol Company expended a certain amount of capital for exploratory work in the 1740s, the indications seem to be that any construction of this nature, such as the long water leat and the engine itself, would be financed by the London Lead Company.

Furthermore, the increased output of lead ore between 1770 and 1773 reflect the supposition that part at least of this ore would have been obtained from below the level of the River Lathkill, in workings which would require pumping to keep them free from water. The complete absence of any ore measurement for the year June 1769 to June 1770 suggests that perhaps the men were employed in other tasks, for example perhaps preparing the leat and engine site. The 'Old Level' (Over Haddon Sough?) previously referred to would no doubt act as a pumpway for the engine.

So much ore was mined at both the Lathkill Dale mines and the Nether Hubbadale Mine between 1767 and 1774, all of which required measuring by the Barmaster, Thomas Roberts, that his salary was increased by the Duke of Devonshire:

3 The Late Eighteenth Century

'for his extraordinary trouble of attendance at Lathkilldale and Hubbadale Mines.'

He received a gratuity of £10 per annum in addition to his salary of £25.32.

During May 1770, the London Lead Company claimed 84 meers of ground on Little Greensorake 'from the wall of Ashford Inclosure to the Head of Lathkill River' (2). On 29th March, 1772 they staked out several meers of ground in Mandale Vein, Smallpenny Vein and others in Bakewell Liberty, whilst the day before they had 'nicked' virtually the same veins within Upper Haddon Liberty. The Company had originally claimed 10 meers of ground in Great Greensort (Greensward) Rake and 8 meers in Little Greensward Rake in Bakewell Liberty in January 1767 (31). Greensa, Greensort and Greensward Rake are various spellings of the same vein. It was being worked in the sixteenth century and through to the latter part of the nineteenth century.

The final entry in the Barmaster's Book relating to the London Lead Company freeing meers on Lathkill Dale Vein is as follows:

'18.5.1773. Received of the Governor and Company at Lathkilldale Mine, one dish of ore for the freeing of the 19th Taker Meer ranging westwardly from the Founder Stake.

Thomas Roberts, Barmaster' (2).

The rate of progress for the driving of Lathkilldale Sough between 1743 and 1764 was about four inches per shift. Mention has already been made of the possible date when it was commenced and also the one measurement of ore in Meadow Place Grange in 1750, There is then silence until 1772, when a Barmaster's entry for the freeing of taker meers by the London Lead Company refers to the 'Lathgilldale Mines and Sough.' The driving done by the London Lead Company between 1764 (by which time the sough must have reached the future site of Bateman's Shafts) and 1773 averaged eight inches per day working two shifts. By the later date the sough had reached the western end of the 19th taker meer.

They continued to free veins until 1774, by which time the Company owned comprehensive holdings on veins extending some distance north and south of the River Lathkill. Their title to Mandale Vein eventually extended from the river in Lathkill Dale, northwestwardly to the Highlow Pipe, a distance of 2¼ miles.

The yield of ore decreased alarmingly after 1773. During 1774 only 244 loads were measured and by 1775 this had dropped to the insignificant figure of 8 loads 6 dishes. It is not the least surprising to find that in July 1776 the Company decided to sell the whole of the Lathkill Dale Mines. On July 25th William Smith and John Taylor the Derbyshire Agents reported to the Company, but the report was not entered in the Court Minute Book. The following observations were passed:

> 'They make the expense of carrying on the said mine to amount to many thousand pounds, but proposed an experiment be tried at the Ridgeway Rake, which may be done at the expense of £10 or thereabouts. It was agreed they do compleat it as soon as possible, which they said would be a fortnight and give the Court an account of the success thereof.
>
> That in the meantime it is the opinion of this Court that the whole of Lathkilldale Mine be sold and that the Agents do enquire for purchasers, and the sale of the Mine Advertised (36).'

The sale was to take place at the Old Bath, Matlock, on the 30th June, 1777. Several other Derbyshire mines were offered at the sale including Watering Close, Millclose, Slack, Gurdall, Bithoms, Davis, Boot Wood and Ballington Wood. The following is relevant to the Lathkill Dale Mines:

> 'Lot 1. Lathgilldale (sic) Title, in the several Liber-ties of Upper Haddon, Bakewell, Sheldon, Monyash, Meadow Place Grange and in the Hundred of High Peak.
>
> For further particulars apply to Mr. William Smith, Winster, or Mr. John Taylor, Stanton, who will shew the premises' (37).

The ore accounts continue without any mention of new ownership until the 17th August, 1779, but the quantities raised were very small, 48 loads being the maximum in any year.

The Hillcarr Sough Company: 1779-1820

The Taylor family and some smaller ventures

On the 17th August, 1779 Samuel Parker, the Agent to the Hillcarr Sough Company, claimed possession

3 The Late Eighteenth Century

The incline entrance to Mandale Mine situated near to, and above, the Engine House.

of the whole of the 'Lathgilldale Tytle,' covering an area extending from Conksbury Bridge westwardly to Lathkill Dale Head, northwardly into Ashford Liberty and southwardly to include a good deal of Meadow Place Grange (2). Within this area, thirty six separate veins with a total length of over twenty miles were listed together with their locations. Unfortunately many of these veins cannot now be traced accurately, although their approximate ranges are known in some instances.

Hillcarr Sough was commenced in June 1766 from the River Derwent north of Darley Dale to drain the mines in the Alport – Youlgreave area, but by 1779 it had not reached Greenfield or Great Shaft south-east of Alport, and did not reach Guy Vein until 1787. By this time it had cost in excess of £32,000 to drive. One finds difficulty in believing that the Sough Company seriously contemplated driving the sough to Lathkill Dale, but an indication of their intentions is provided by the following entry:

3 The Late Eighteenth Century

Short trail level by the path near to Mandale Sough tail.

3 The Late Eighteenth Century

'8.10.1781... gave John Taylor 33 meers of ground on the Pasture Rake in Upper Haddon Liberty ... the same being nicked for want of workmanship, and there is a reserve made in the gift, that is, if the present partnership of Hill Carr Sough Levell brings or causes the said Sough Levell to be wrought or brought forward into the said Pasture Rake, the said partnership is to have the benefit and use of the bottom or under the said John Taylor's Works, the same as if this gift had not been done or acted' (2).

Nothing very much was done because Hillcarr Sough certainly never came anywhere near to this part of Lathkill Dale. A branch level called Danger Level was driven during the 1860s along Windy Arbour Vein, parallel to the River Lathkill between Alport and Coalpit Bridge, but was abandoned with its forefield many thousands of feet short of the portion of Lathkill Dale discussed above (38).

On 25th April 1782, twenty-five meers of ground were nicked along Lathkill Dale Vein and given to Philip Taylor 'for want of workmanship.' At this date the Lathkilldale Sough Forefield Shaft was three meers below the west end of the ground, twenty two meers west of the boundary between Meadow Place Grange and Upper Haddon at Bateman's House. Two meers were reserved for Joseph Taylor to have the 'Flatt or Flott ore' in the old hillocks (2).

Working during the next twenty years was on a very limited scale, although Mandale Vein, Pasture Rake, Gank Hole Vein and others were all freed from time to time. One rather interesting development at this time was the intended driving of a supposed level from the upper reaches of the Dale to the Magpie Mine at Sheldon. This apparently unlikely story has been handed down by generations of Monyash miners as a positive intention (39). The very short level, now known as Ricklow Mine, was started by a Monyash miner, Isaac Beresford and his sons, the original name being 'Beresfords' Level' or 'Barsett's Cut.' The initials 'I. B.' and the date 1787 are carved at the level entrance.

3 The Late Eighteenth Century

The tail of Mandale Sough in summer. Contrast this with the colour photograph.

4 The Mandale Company 1797-1839

The Driving of the Sough and Two Large Ore Strikes

Mandale Sough was started, according to one source in 1797, or perhaps in 1798, and was over twenty years in reaching its prime objective, Mandale Mine. The main workings in Mandale Rake were reached in 1820, but the level was extended after this date, finally reaching Pasture Shaft, or a little beyond.

On 1st June 1798, Thomas Woodruff was given possession of 60 meers of ground in Mandale Vein and 45 meers in Pasture Rake (2). One could speculate whether Mandale Sough was started shortly after this date. There were other partners in the venture at this time, because in February 1800 Joseph Taylor, a miner of Over Haddon, sold a 1/48th share of Mandale Mine to Sarah Roberts of Taddington for £2. By November 1800 Thomas Woodruff was given

> 'all those parts or shares of a mine called Mandale which belonged to Mr. Humphrey Winchester, by a verdict returned at the Barmote Court, 10th October, 1800.'

On the same day he was given all the shares formerly held by James Shemwell and the shares belonging to Miss Frances Goodwin were given to Samuel Hollis, all by verdicts returned at the Barmote Court. A note signed by Woodruff in 1800 stated:

> 'Robert Tonge is entitled to a 1/24th share of a mine called Mandale Sough and Mine from the first when it was set on work by Thomas Woodruff'.

Woodruff was agent to the mine and sough until 1808.

The sough was 700 yards long in May 1807 at which time it was intended to drive it to 'Mr. Winchesters Shaft on the More,' a total distance of 1450 yards or a

4 The Mandale Company 1797-1820

A small stope in Mandale Mine.

4 The Mandale Company 1797-1820

little beyond the site of the later 'Top Engine Shaft' (40).

Reckoning books (41) covering the period from 1808 to 1836 indicate that between July 1808 and September 1813, 252 yards were driven, and a letter written in December 1813 states that 200 yards yet remained. The extent of driving between 1807 and July 1808 is not known, but apparently the original objective, i.e. Winchester's Shaft, had not been reached by 1820. From the available evidence it seems probable that the sough would then be very near to reaching the area of much worked ground between the 'shaft in Duke's Piece' and the 'old Mr. Winchester Shaft' The position of the shaft in Duke's Piece is shown on a section of Mandale Mine.

During 1814 there was wage work in the Old Level, as distinct from the Sough Level, and in 1826 six men were paid £4 5s. for opening 15 fathoms in the Old Level. Whether in fact this 'Old Level' was the original Pasture Rake Sough of the 1730s is not known, but seems to be a possibility. One account states it was only 1$^{1}/_{2}$ yards above the Mandale Sough, so that it was presumably an old drainage level at river contour.

The available reckoning books (41) commence in July 1808 and during that year 31 fathoms were driven in the Sough Level costing £103 15s. Strangely enough nothing seems to have been done in the Level during 1809, but 8 men were paid a total of £5 'for sinking a Levile shaft at the Levile mouth, 5 fathom.' Between January 1810 and March 1811 a further 20 fathoms were driven at a cost of £85 8s. From then until February 1812 another 32 fathoms in the sough forefield cost £143.

The level was inspected regularly and William Wager who had taken over as agent in 1808 paid £1 to 'old Samuel Hollows for going throw the Levile at sundry times.' Most of the sough was driven by a gang of six or eight men supervised by Samuel Hollis (or Hollow, or Hollows). There was an 'old' Samuel Hollows, Samuel Hollows (or Hollis) Senior, and his two sons Samuel and George. George was killed by a fall of stone in the mine in August 1836, an Inquest being held by the Barmote Court on the 2nd September. Only after the passing of the High Peak Mineral Courts Act in 1851 was a Coroner allowed to hold an Inquest over a death in the mines.

The Hollis family of Youlgreave were associated with lead mining for at least two centuries. Tithe ore was being paid by a Denis Hollis

Section along Mandale Rake

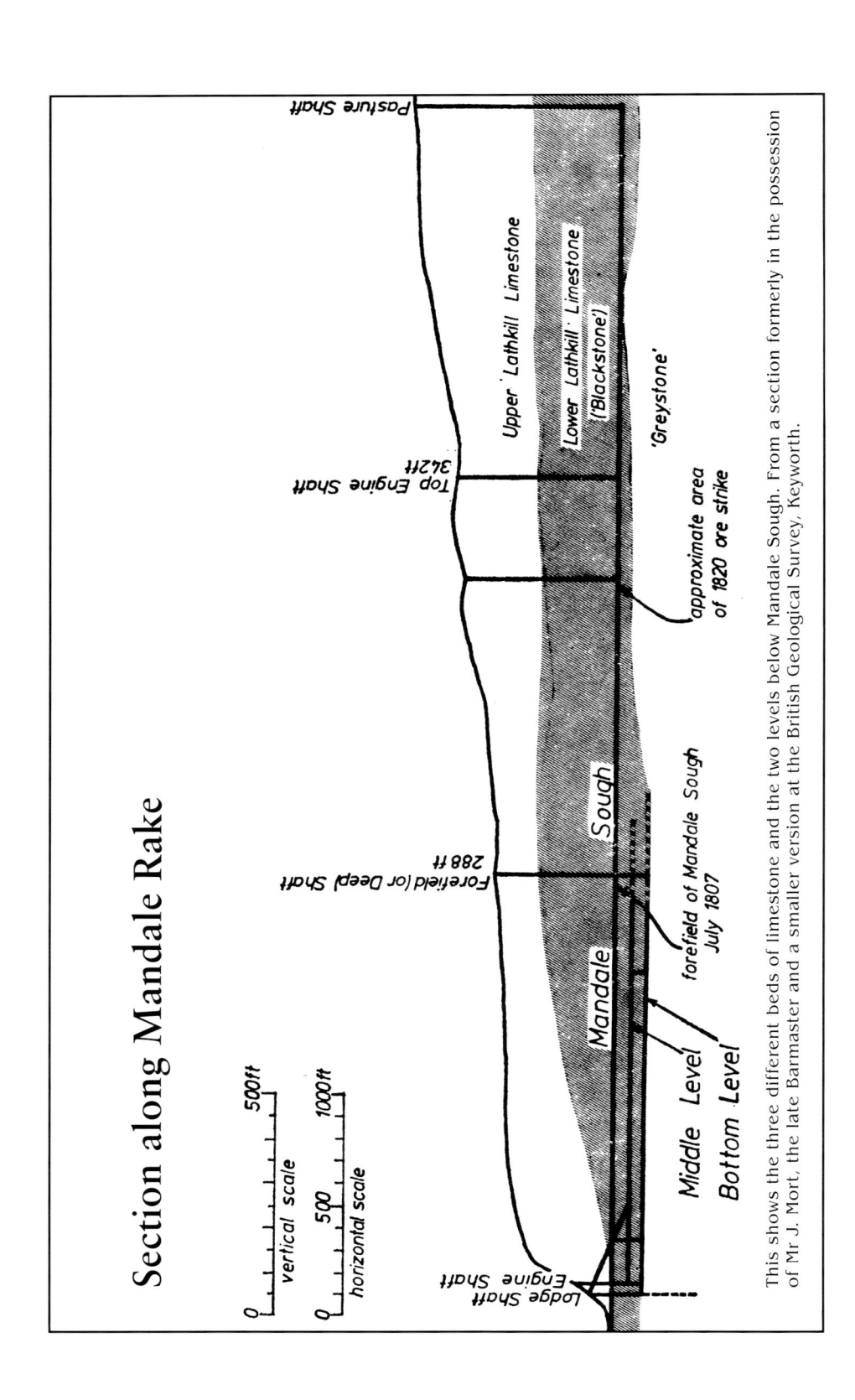

vertical scale: 500ft
horizontal scale: 500, 1000ft

Lodge Engine Shaft

Bottom Level
Middle Level

Mandale Sough

Forefield (or Deep) Shaft 288ft

forefield of Mandale Sough July 1807

Top Engine Shaft 342ft

Pasture Shaft

Upper 'Lathkill' Limestone
Lower 'Lathkill' Limestone ('Blackstone')
'Greystone'

approximate area of 1820 ore strike

This shows the three different beds of limestone and the two levels below Mandale Sough. From a section formerly in the possession of Mr J. Mort, the late Barmaster and a smaller version at the British Geological Survey, Keyworth.

4 The Mandale Company 1797-1820

in Youlgreave Liberty in 1669, (42) and Samuel Hollis was mining as far away as Ible in 1735 in partnership with three other miners from the village. Several members of the family served on the Grand Jury of the Barmote Court for Youlgreave Liberty between 1770 and 1808; by 1806 both Samuel senior and Samuel junior were Jurors (33). Many other Mandale miners at this period served on this Jury alongside the Hollis's including William Ragg, Joseph Thompson and members of the Brassington, Twigg and Buxton families.

Between 1808 and 1819 two gangs of men worked the sough earning 2s. per day on wage work, somewhat less when on bargain. During these eleven years only 27 men are mentioned in the reckoning book (54). Beside the Hollis family, other miners employed at this time were James and John Taylor, both of Over Haddon. Joseph Thompson, John Rooper, William Wildgoose and William Ragg appear to have worked intermittently at the mine. Wildgoose was probably from Sheldon, the others from Youlgreave.

From February 1812 until May 1813 only 23 fathoms were driven at a cost of £153 and it is interesting to note how the cost of the bargains was continually rising. In 1808 payments between £3 and £3 10s. per fathom were made; by 1810 the prices fluctuated between £3 and £8 per fathom and by 1812-13 a further rise from £4 to £8 per fathom, with one payment of £10 and one of £12. Very often the miners found the rock so hard that increased payments were made to them over and above their contracted bargains:

> 'September, 1810. Samuel Hollows and partners, 8 men for Driving in the forefield of the Levil, 10 fathom at £3.3.0 fathom, £31.10.0.
>
> The same bargin Turning out so hard, Mr. Wakefield agreed I should give them more. £8.8.0.'

This entry also indicates that John Wakefield of Kendal was already a major influence at the mine. He later became principal shareholder, and Chairman of the Mandale Mining Company possibly as early as 1839.

A very illuminating letter was written in December 1813 by George Cantrell, the clerk at the Ecton Copper Mines, to W. E. Sheffield, the agent at the Whiston Copper Works. Cantrell had been requested by Sheffield to view and report on the condition and prospects of the mine. He was:

> 'astonished to find a Level of that sort cut out in so little compass, the distance

4 The Mandale Company 1797-1820

they have to drive the Level before they expect to meet with the Mandale Vein, the object in view is about 200 yards ... young Mr. Wager, the Agent's son reckons upon driving 4 or 5 yards a week with a company of 8 men, and which appears to me rather extraordinary why he should state it at so little for its my opinion were there a Level of the same kind going forward at Ecton, the same number of men must certainly drive 8 or 10 yards per week. There is no opportunity of seeing the old workings of the Mandale 'till the Level is completed ...(43).'

Before the large ore strike of 1820 was located, a further 241 fathoms were driven, and as before stated the sough forefield would then be approaching the eastern fence of the enclosure of the much worked area of ground which contains the old Founder Shaft, Top Engine Shaft and Winchester's Shaft.

Four men began sinking a new engine shaft in September 1814, and this had reached a depth of fifty fathoms exactly four years later. Samuel Hollows supervised this sinking. A new wind shaft was also sunk at this time and reached a depth of fifty-one fathoms. The Engine Shaft was most probably the shaft later referred to as 'the shaft in Duke's piece', but there is a slight possibility that it could have been the Top Engine Shaft. The former shaft is now run-in, and until about 1980-1990 the site was marked by a grassed over gin-circle, thirty feet in diameter. The Mandale Company paid £10 for an old engine from 'Greensalrake' mine and this would be a horse-gin to work on the newly completed shaft.

Little ore was mined, in fact only £56 5s. 11d. worth of galena was produced from 1808 until 1820, resulting in a nett loss to the mine proprietors of £3,407 12s. There then followed a remarkable ore strike, the sort of 'bonanza' old metal miners always dream about, are always confident exists, and most certainly breeds the optimism so inherent in metal mining.

A big ore strike

During a period of 26 weeks, from 20th May to 18th November 1820, lead ore to a gross value of £1822 1s. 1d. was obtained on which a nett profit of £1,155 2s. 2½d. was made. Samuel Hollis, who had worked at the mine during the long years of driving the sough, produced 152 loads of ore for which he was paid

14s. per load. His wages averaged £2 8s. per week, until the re-letting of copes in September 1820, when he obviously missed obtaining an advantageous cope bargain and reverted back to wage work. By 1823 he was again getting ore, this time leading a gang of 16 to 20 men, but he averaged only 7s. to 11s. 6d. per week.

There was, as might be expected, a large influx of new miners at this time. One of them, Critchlow Brockley, and a gang of eight men measured 244 loads at 18s. per load. These eight miners each averaged £2 15s. over a 10 week period, but their profits fell during the next 37 weeks and Brockley's name does not re-appear after 1821. He was a Magpie miner and probably came from Sheldon.

A second ore strike was made in 1823, when lead ore to the value of £980 14s. 5d. was raised between 11th April and 6th September with a resultant profit of £585 6s. 10d. Ninety-one different names appear at one time or another between 1820 and 1824; gangs of between 4 and 12 men are recorded in 1820-21, and by 1823 larger gangs of 16 to 20 men were raising ore.

Letters written by John Wakefield of Kendal, the principal shareholder (he owned 13/24ths of Mandale Mine by 1814), to William Wager the agent, show that the bulk of the ore obtained during these 'strikes' came from wide 'bellies' or 'boxes' in the limestone (1). A section of the workings shows that in 1820 the forefield of the sough would be in the 'Blackstone', the miners' term for a dark coloured, bituminous limestone, which was subsequently mined and quarried higher up the Dale (8). When polished it became the well known Derbyshire 'Black Marble.' However these were not much more than trials owing to an excess of chert.

Running at a Loss

The sough was always a constant worry for it was very small, being only 3ft. 6ins. high and 1ft. 6ins. wide, and parts were driven through very bad, shattered ground. Wakefield, writing to Wager in July 1826, noted:

> 'Gantley gives a poor report of the state of our present sough. It must be attended to and kept safe until the lower one makes it useless.'

Eventually there were two lower levels in the mine, but apparently about this time a branch sough was driven up Mandale Vein from Lathkilldale Sough. How far this was eventually driven is not known, but the above letter also gives a further reference:

4 The Mandale Company 1797-1820

'From an account I have received from John Gantley there appears a much greater fall between the two suff mouths than I had reckoned on. He makes the difference 40' 6". How the old Company under Woodruff neglected availing themselves of so great an advantage seems to me most unaccountable.'

A further letter dated the 27th October, 1826 goes on:

'... do let me hear ... if any progress has been made in the Agreement with Bateman and Alsop to commence operations in Lathgil Dale.'

The 'two suff mouths' were presumably Lathkilldale and Mandale Soughs. A branch sough out of the former would allow deeper drainage, and, as no mechanical pumping was attempted until 1840, such a level would offer obvious advantages. There is an isolated reference in 1854 to the 'Mandale and Lathkill Co's Deep Level' (9).

Shortly after the second of the two large ore strikes at Mandale Mine activity quietened considerably. During the years from 1820 to 1825 as many as sixty men and more were employed at a single reckoning, both on cope and on wage and bargain work. Disastrously for the Company they were not again destined to locate lead ore in quantity as they had done in the early twenties. With the exception of 1827-28, losses were recorded each year until 1836 when the old company virtually ceased. A small profit was made in 1827, whilst in 1828 ore to the value of £638 18s. 8d. was mined with a resultant nett profit of £315 13s. 4d. After 1826, the work force contracted considerably and only nine men are mentioned between 1827 and 1836. Payment for bargains and wage work was kept to a minimum and only £58 was paid in 1827 in contrast to £404 18s. 6d in 1824. These figures do not include payment to miners for mining lead on cope (41, 54).

Despite two good returns of ore, from July 1808 to December 1836, the proprietors expended £9,167 16s. on the driving of the sough and maintenance of it as well as other work in the mine, yet only mined ore to a value of £6,544 19s. For example, £450 was paid for sludging and repairing the sough, exclusive of timbering, walling, and cost of materials.

The operations from 1808 until 1836 had cost the proprietors a nett loss of £2,622 17s., but they continued to entertain high hopes for the eventual success of the mine.

5 Power to Drain the Mines:
The Lathkill Dale and Mandale Companies

1. Lathkill Dale Mine 1825-42

During these years of activity at Mandale Mine little is known about Lathkill Dale Mine. The Taylor family retained an interest in the title until the 29th of September 1825 when Thomas Bateman and John Alsop bought 25 meers of ground in Lathkill Dale Vein for £30 from James and John Taylor (44). The Taylors however retained the right to search the old hillocks for what they could find. The 25 meers were those given to Philip Taylor in the gift of 1782.

On the 3rd November, Bateman and Alsop were given 18 meers of ground as taker meers at their old possessions on Lathkill Dale Vein, i.e. beyond the existing twenty five meers, ranging west to Mr. Thomas Finney's Plantation on the top of the hill. They were given part of Greensarake Vein, and other sundry veins ranging onto Haddon Moor. At this period their agent appears to have been Abraham Haywood of Middleton-by-Youlgreave. On the same day they obtained 32 meers of ground also on Lathkill Dale Vein, this time ranging eastwardly through Meadow Place Wood to Upper Haddon Corn Mill. This part of the vein is in Hartington and The Granges Liberty (45).

Shortly afterwards, a plan, dated 1826, was drawn showing the range of Lathkill Dale Vein, the course of an old water leat and the position of a former engine. The possible significance of this latter piece of machinery has already been discussed, the plan being also important in showing the line of what is termed 'the proposed water course.' This water course or leat was to be constructed from a little east of the later named Carter's Mill and was to terminate slightly downstream from what

5 Power to Drain the Mines

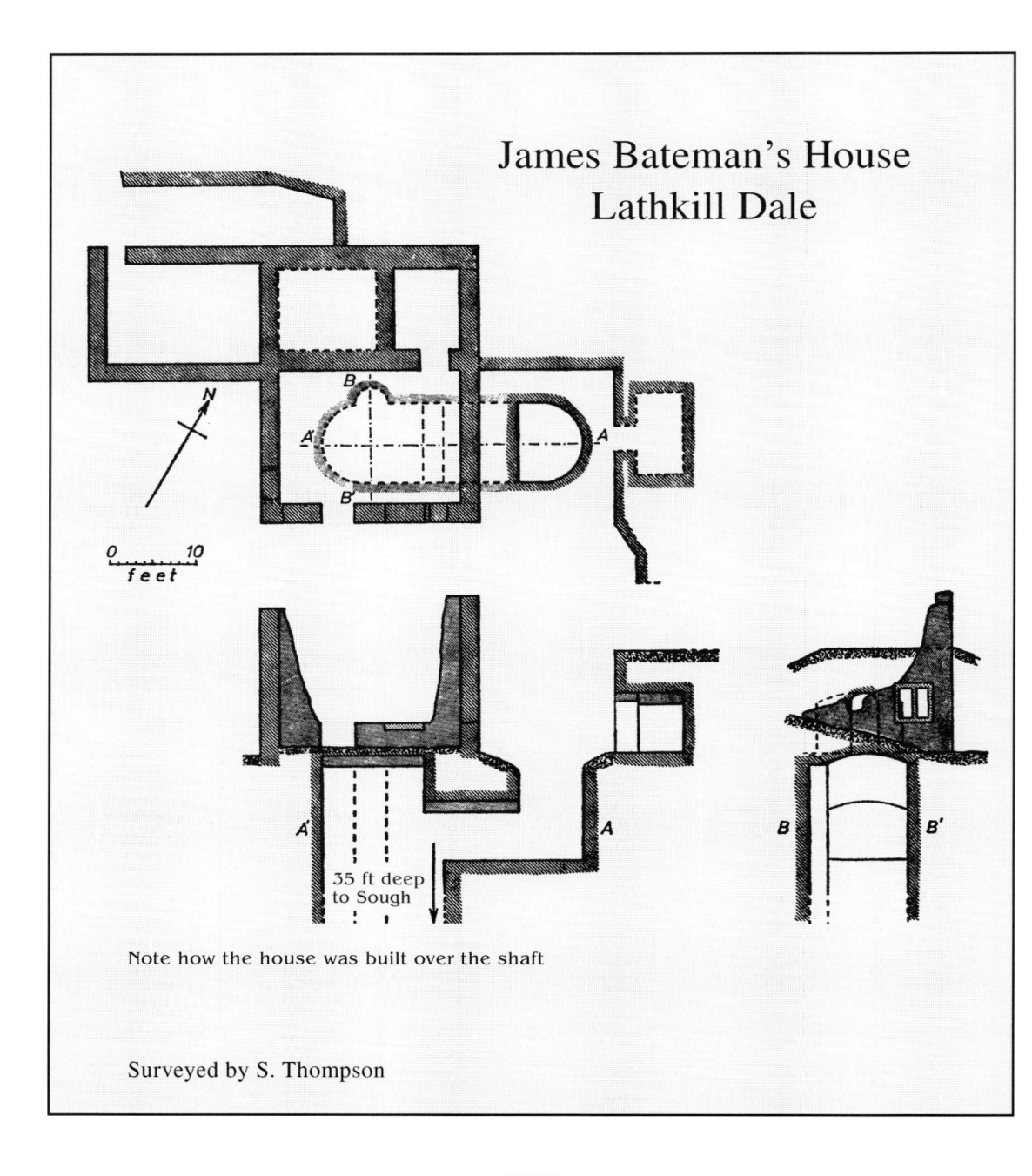

James Bateman's House
Lathkill Dale

0 — 10 feet

35 ft deep to Sough

Note how the house was built over the shaft

Surveyed by S. Thompson

5 Power to Drain the Mines

was to become the site of James Bateman's House. A second plan, almost identical, is attached to a lengthy Agreement dated 27th October 1827, in which Lord Melbourne, the owner of the Estate, granted permission to Bateman and Alsop to erect:

> 'one or more water wheel or wheels or other Engine or machinery or work for raising and drawing water'.

The lease was for 21 years at £25 per annum, the first payment being due in March 1828 (46). No indications are given on either of the plans as to where any machinery was to be erected, neither is the type of machinery specified.

An entry recorded in a Hartington and the Granges Barmaster's Book in April 1829 noted:

> 'put Francis Nuttall into possession, (for the use of Thomas Bateman and Co.) of four meers of ground from Larkhill River in a vein ranging southwardly up the wood by the New Shaft. Also put the said Francis Nuttall into possession of fifty-six meers of ground in Larkhill Vein, commencing at Larkhill (sic) River and ranging westwardly to near One Ash House' (47).

These meers were probably never worked. They formed a southwardly continuation of Lathkill Dale Vein from the south side of the river at Carter's Mill.

There is no proof, but probably the 'New Shaft' is one of the shafts at Bateman's House. The indications, such as they are, on both the 1826 and 1827 plans seems to be that the pumping arrangements served by the 'proposed watercourse' were to be sited somewhere in this vicinity and not at the site of the later, very large water wheel.

An Unusual Engine

One possibility, which seems to have a reasonable amount of evidence, although there is no definite proof, is that a Disc Engine patented by the Dakeyne brothers (flax spinners of Two Dales) in 1830, was intended to work at the shaft beneath Bateman's House. This engine was described in minute detail by Glover (48). Essentially, it consisted of a circular, flat outer casing fitted around the equator of a large central globe. Free to move inside the circular casing was an annular ring. An axial spindle connected through either a crank and crankshaft, or a gear wheel arrangement to the pump rods.

(cont'd on page 58)

Above: Bateman's House 1950. **Below:** The house today after being robbed of the lintels.

Batemans house in July 1958. The family were presumably oblivious of the shaft beneath the house, especially with the girl playing around the back.

5 Power to Drain the Mines

Falling water was forced under pressure around the disc or annular ring within the casing causing the whole globe to tilt rhythmically, but it did not rotate, thus causing a precessional movement in the spindle. This in turn rotated the crank or gear wheel, which connected through further gearing or cranks and thereby activated the pump rods. The engine could conveniently be likened to the planet Saturn with its rings.

Glover stated that such an engine, of approximately 140 horse power and to work under a head of 66 feet of water, was in the course of erection at that time on an extensive scale for Messrs. Bateman and Alsop at their mine near Alport. The location appears to be the one damning factor against this engine being situated in Lathkill Dale. One would imagine Glover was so familiar with Derbyshire that he surely would have differentiated between Alport and Lathkill Dale, but in all other respects, including the 66 feet head of water from the leat to the sough, facts fit. The engine was therefore almost an elementary turbine, but as Dr. P. Strange has pointed out, the engineering difficulties of working this engine would have been pretty formidable. A head of

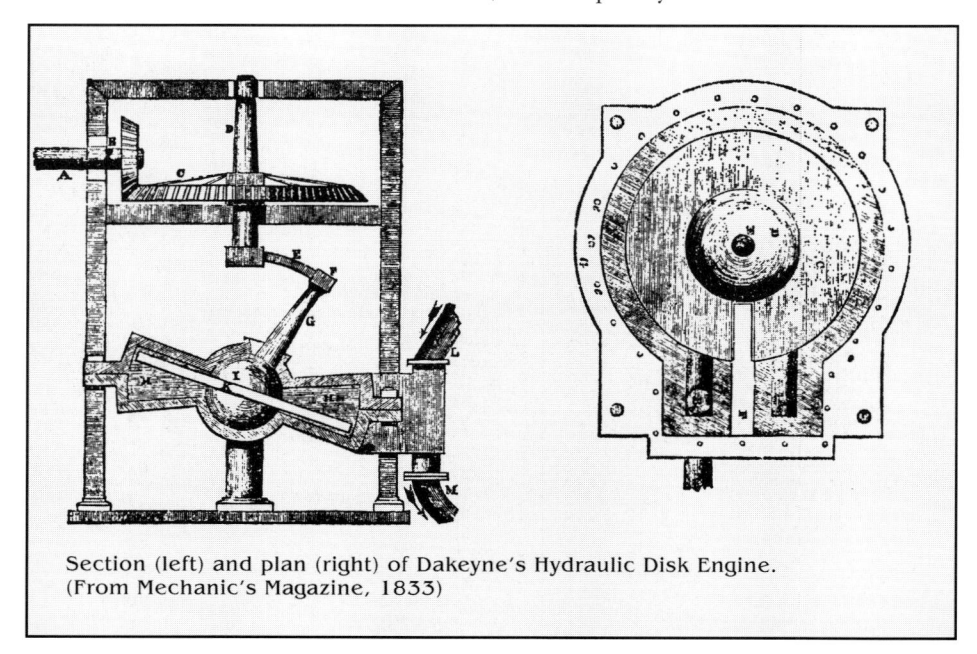

Section (left) and plan (right) of Dakeyne's Hydraulic Disk Engine.
(From Mechanic's Magazine, 1833)

5 Power to Drain the Mines

The approach to the shaft. Above the explorer's head is the underside of the house floor.

Above: The shaft below Bateman's House. The masonry is particularly fine.

water of 66 feet would give a working water pressure of 28.5 lbs./in² (1.8 tons/sq ft).

There are two further references to this Dakeyne engine which are of some relevance. The first is contained in the *Derbyshire Courier* for the 22nd January, 1831, and is also quoted in *The Smiths of Chesterfield* by Philip Robinson, published in 1957:

'a heavy casting of upwards of 7 tons was recently run at the Adelphi Iron Works of Benjamin Smith and Co., Duckmanton, being the upper part . . . of the vast orbicular case. . . to Dakeyne's patent hydraulic engine which is erecting by the inventors. . . for Thomas Bateman of Middleton and John Alsop of Lea to be applied to the purpose of

5 Power to Drain the Mines

raising water by the power on continuous descending columns of that element at their mine near Alport. This original and powerful machine is estimated at upwards of 130 horse power, with 66 feet vertical fall of water, giving rotary motion to the machinery' (49).

A second reference, contained in the *Mechanics' Magazine* for the month of January 1833, includes a reasonably detailed drawing of the engine with the added comments by a correspondent calling himself 'Galena':

> 'an engine on their (Dakeyne's) plan, of very considerable power, is now actually at work on a lead-mine belonging to Messrs. Alsop & Co' (50).

The editor of the magazine commented that Dakeyne's patent specification was one of the worst that he had ever seen, being 'a perfect smother of outlandish jargon.'

At the bottom of the shaft beneath Bateman's House is a large excavation, presumably made to house some piece of machinery. The location of this large diameter shaft, immediately beneath a building, has always been somewhat puzzling, but S. Thompson (pers. comm) thought that it might indicate work of an experimental nature. If the Dakeyne engine was installed at the bottom of this shaft, it is quite conceivable, it being a new invention, that the interested parties would wish to conceal it, hence the protective building over the shaft. From examination of an old photograph it appears that the second shaft was also concealed beneath a building.

A notice dated 14th August, 1830 invited tenders for the letting of bargains for: 'The sinking of a Large Shaft or Lodge at the Lathkiln-Dale Mine' (51). Most probably this was the Dakeyne pumping shaft.

From 1834 until 1842, when Bateman and Alsop ceased working the mine, the reckoning book is available (52). A further Lodge was sunk in 1834-35, most probably this one being the pumping shaft for the large water-wheel. During 1835 and 1836 an enormous amount of work was done in preparation for the erection of this wheel. There is some evidence to suggest that two or even more water wheels operated on Lathkill Dale Vein at this period:

> 'The drainage of the river is chiefly owing to extensive mining operations between 1830 and 1850 in seeking for lead ore, when

5 Power to Drain the Mines

Looking down Bateman's shaft to the water in the sough below.

several water wheels were employed in pumping water from the workings'.

Two wheels were offered for sale at Lathkill Dale Mine in 1847, consisting of the great wheel and a smaller one of 11 feet diameter.

The Huge 52-Feet Diameter Water Wheel

The leat, which presumably had been originally constructed to serve the pumping arrangements at Bateman's House, was widened and 474 man-days were spent coating the bottom of it with 509 loads of clay puddle. In March 1836 the sum of £30 was spent 'making room for the wheel.'

In the September, Benjamin Smith's Duckmanton Foundry, near Chesterfield, was paid £537 for castings. Richard Page was paid £30 'for his attendance and planning from the commencement'. Page was the Resident Engineer to the Alport Mining Company, but he appears to have acted in an advisory capacity to the Lathkill Dale Company. He was involved not only in the erection of the wheel, but also earlier in 1830, when he was involved in the sinking of the first large Lodge shaft, and thus by implication, in the erection of the Dakeyne Disc Engine.

The available evidence seems to indicate that the large wheel was erected during the latter months of 1836, at which time payments commence for 'tenting (attending to) the wheel, Sundays.' The wheel was 52 feet in diameter and 9 feet across the breast and was described in 1846 as being 'the largest except one in the Kingdom' (53). It was rated at 140 horse power, lifted 4,000 gallons per minute from 120 feet beneath Lathkilldale Sough, via six sets of pumps, each 18 inches in diameter. The wheel had been removed by 1861.

Workers at the Mine

From 1834 until large scale working ceased in 1842, a total of 124 miners were employed at the mine, although few stayed for any length of time. This drifting labour force was an integral feature of Derbyshire lead mining. As an example, during the years previous to the rich ore strike at the Nether Hubbadale Mine, Taddington in 1767, over twenty mines were in production within a four mile radius. This was reduced to a mere four between 1770 and 1774, during which time only Nether Hubbadale Mine (until 1772) and Lathkill Dale Mine were major producers. After the decline of both these mines, the smaller

5 Power to Drain the Mines

ventures increased in numbers again (32).

There was a rapid intake of labour at both Hubbadale and Lathkill Dale mines when they 'cut rich,' but only at the expense of the smaller producers. Similarly, there was a large influx of miners at the Mandale Mine in the 1820s. Undoubtedly, as in previous centuries, many of the small mines were only worked on a part time basis, principally in conjunction with farming, with few men employed at each. During the 1830s and 40s, the men on day work at Lathkill Dale Mine earned 2s. per day, but many of them did a variety of work, including mining the lead ore on cope bargains, driving levels, and sundry work both above and below ground. A gang of eight men received in the order of 35s. per load of dressed ore.

Many of the miners came from Youlgreave, at least forty in fact, while only thirteen came from Over Haddon. Over forty others cannot be traced, whilst a few came from more remote villages such as Monyash. Several old miners paths are known between Youlgreave and Lathkill Dale, and at least one is said to be paved, the slabs still in position although covered by grass. The chief miner was John Knowles who lived at Cauldwell End, Youlgreave. He was the leader of a gang of men between seven and nine in number and all types of work were undertaken. This illustrates how versatile these miners had to be: they not only mined the lead and drove levels, but sank and walled shafts and many jobs on the surface were tackled. Later, during the 1840s, he became the agent at Wheels Rake Mine, Alport. Christopher Brassington who had worked at Mandale Mine also worked here for a considerable period (54).

Coal was supplied from Stubley, near Dronfield; this Colliery also supplied coal to the Fairbairn pumping engine at Watergrove Mine, Wardlow Mires. About ten kibbles were used every year, each costing 3s. 6d; candles were 6d. per lb. From September 1839 until June 1842 a full time blacksmith was employed earning 2s. 10d. a day. A smith was a most important man, a local one generally serving a number of smaller mines, but most large mines had their own man. The sharpening and tempering of drills and wedges was of obvious importance, quite apart from the maintenance of machinery.

From 1832 until his death in 1836, John Sheldon was the Company's agent, being succeeded by perhaps the best known of the mine

Shaft beneath James Bateman's House, showing backed-up water threatening the stonework.

Tail of Mandale Sough.

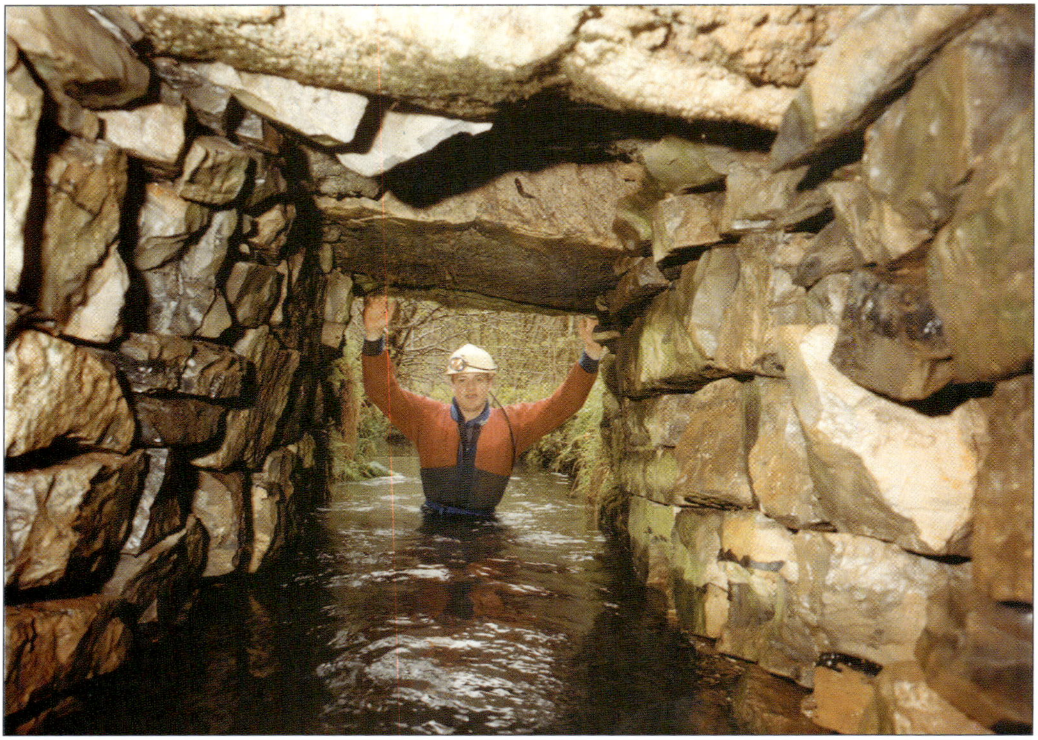

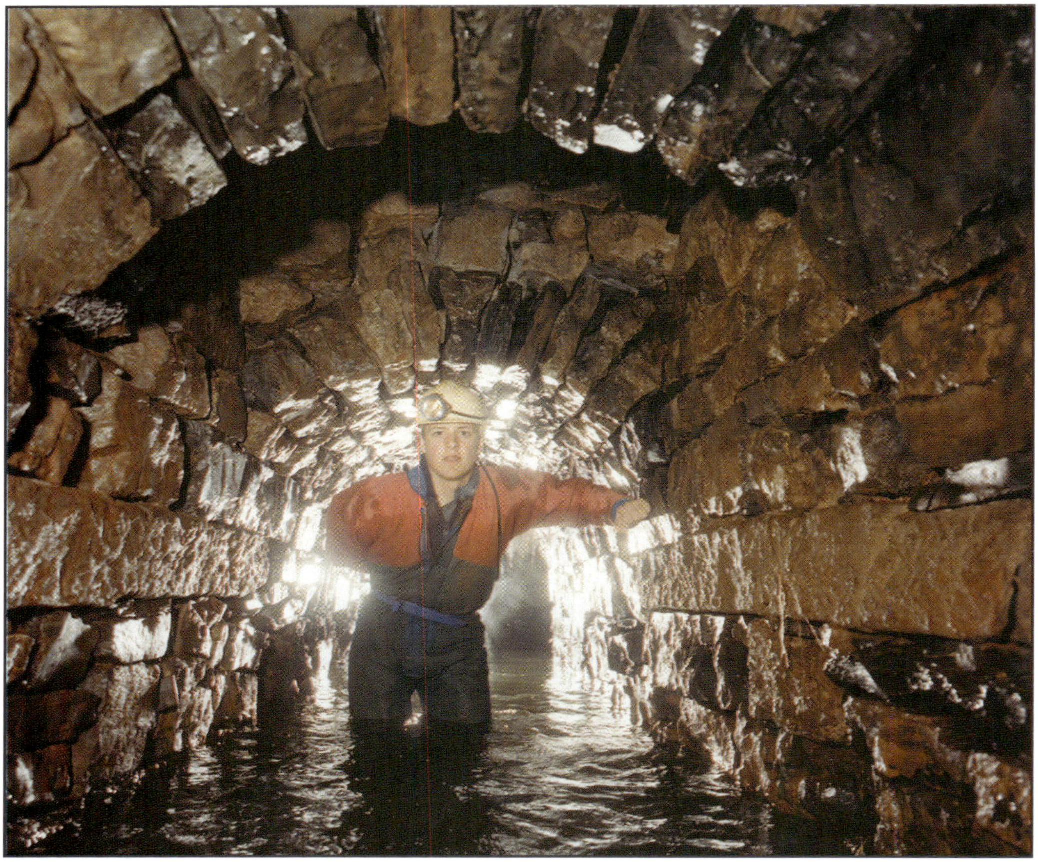

Above: The incline in Mandale Mine.

Right: Exploring Mandale Sough.

Top Left: Mandale Sough just inside from the tail.

Bottom Left: Mandale Sough with shaft up to the surface (see page 85).

Cross cuts from Mandale Sough.

Above: Exploring workings in Mandale Mine.

Left: A stope in Mandale Mine with a small rider adjacent to the explorer's left elbow.

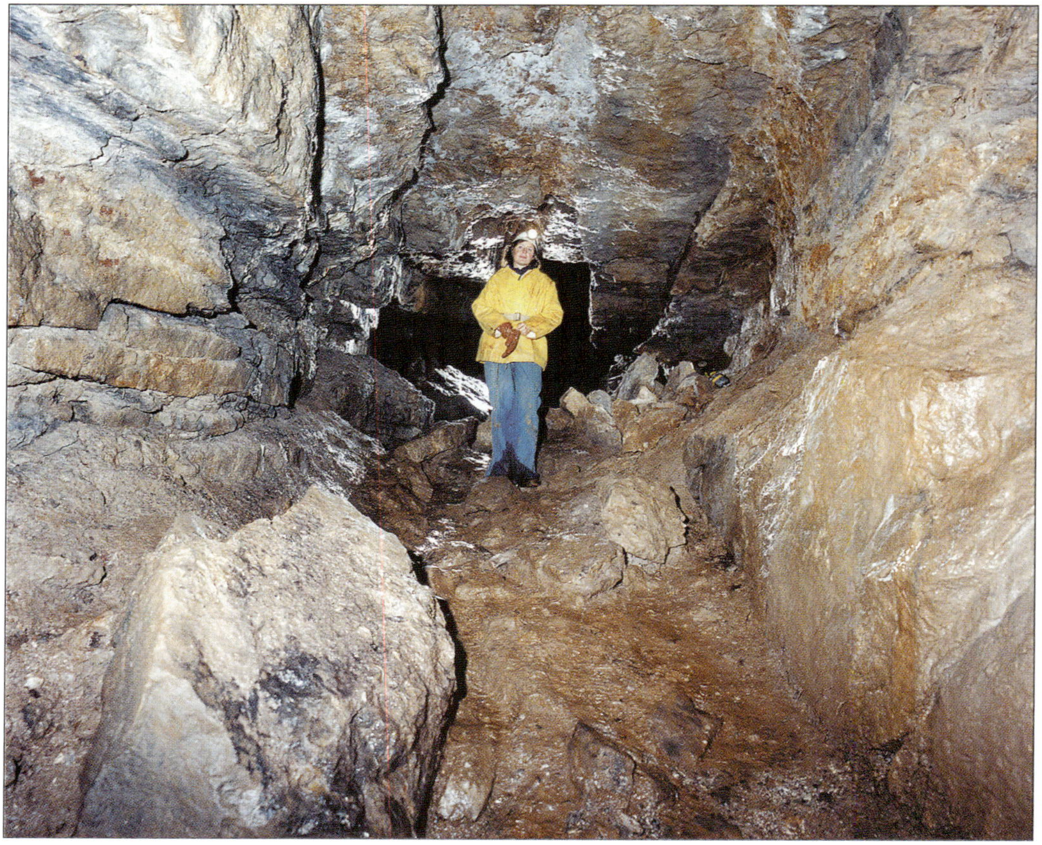

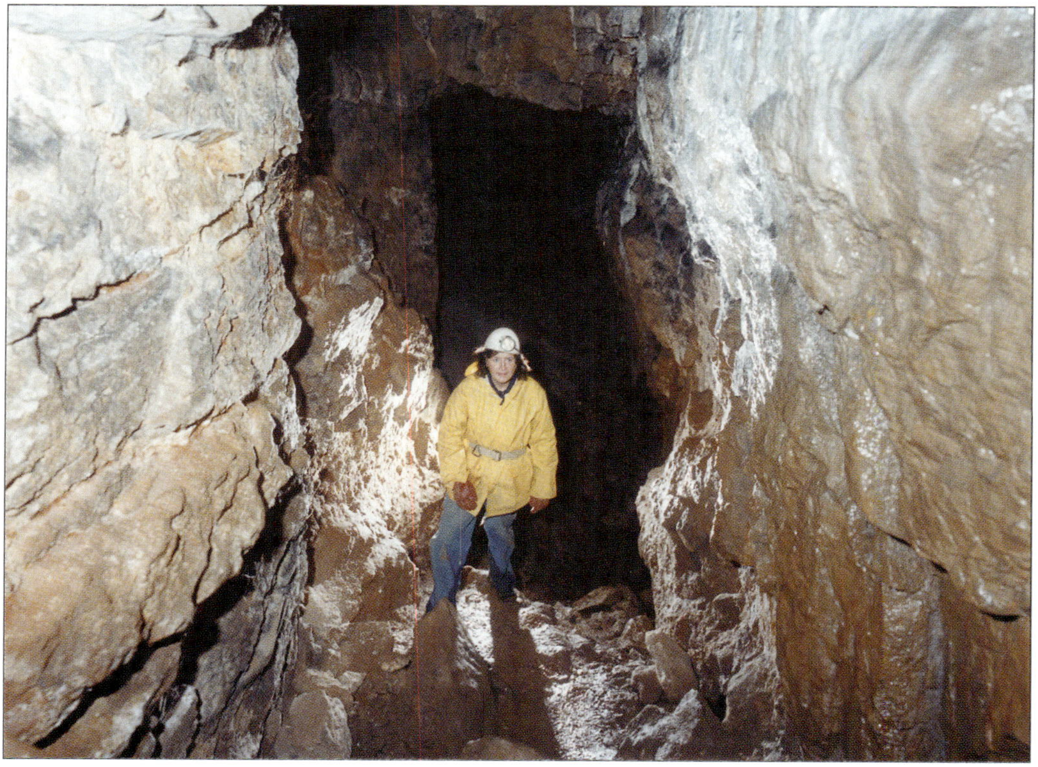

Above and Inset: The narrow width of Sideway Level shows up well here.

Top and Bottom Left: Two further views of the workings in Mandale Mine.

Two views exploring the shaft in Sideway Mine.

agents, James Bateman. He was born in Middleton-by-Wirksworth in 1797, died in 1869 and is buried in Bakewell Churchyard. He received a salary of £1 per week, living in the house which even today bears his name. Later, during the 1840s he was agent to the Mandale Mining Company. He worked Lathkill Dale Mine and Mandale Mine after both had been abandoned by the larger companies. The last recorded ore measurement against his name was in 1863 for the above mines, with one isolated measurement at Small-penny Vein in 1864.

The Company employed a full time engineer. From 1835 to 1840 the position was filled by Adam Killer, then a man in his late sixties. He received a wage of 4s. per day, and was succeeded in 1840 by James Brooks at a similar salary. Brooks remained until the mine closed, by which time (May 1842) he was also engineer to the Mandale Mining Company. He was a young man about 42 years old in 1840 and was the nephew of Richard Page, the former resident engineer to the Alport Mining Co. (54).

The End of Activity

The Lathkill Dale Company, during the time they operated the mine, seldom raised enough ore to cover the expense of working. There were occasional short periods of increased ore production, for example between 1834 and 1839, presumably due to working the vein at depth after 1836 by means of drainage afforded by the water wheel. During 1838, 101 loads of ore were mined from Lathkill Dale Vein in Hartington and the Granges Liberty, the only year that ore was measured from the vein in that Liberty (47).

Lathkill Dale Vein was worked chiefly in Over Haddon Liberty from the river at Bateman's shafts westwardly, but one could wonder whether the hade of the vein changed from north to south, thereby taking the lower levels of the vein to the south side of the river and thus into Hartington Liberty. Possessions were also set in Hartington Liberty in April 1829 for:

> 'fifty-six meers of ground in Larkhill Vein, commencing at Larkhill River and ranging westwardly to near One Ash House' (45).

Despite this, these meers probably remained unworked. The mine ceased large scale working in June 1842, but small amounts of ore were measured in the Company's name until 1851. The wheel was offered for sale on the 16th September,

Above: The breast wall of the launder which carried water over the river to the 'great wheel' at Lathkill Dale Mine. **Below:** The remains of the pit which housed the 'great wheel'.

5 Power to Drain the Mines

1847, together with the rest of the mining property, including a small water wheel of 11 feet diameter and 250 yards of foreign deal troughs or launders (55). There are various small descriptions of Lathkill Dale Mine in several nineteenth century guide books. *Bagshaw's Directory* of 1846 commented:

> 'The township (Over Haddon) is noted for its mines, ... the Lathkill worked by Messrs. Alsop, Taylor and Co., which is a very extensive mine but for the last three years has been disused, owing to it being overflowed with water though an overshot water wheel of 52 feet diameter had been employed (53).

Croston, in *On Foot Through the Peak*, published in 1862 said that a mine in Lathkill Dale was worked by Messrs. Alsop, Taylor and Co. for a period of about 20 years, since when it had been overflowed with water. His description is such that he can only be referring to Lathkill Dale Mine.

William Adam in 1861 spoke of the 'leviathan wheel' fifty feet in diameter, but by that date removed from the Dale, the scene being one of utter desolation:

'the miners coes thrown down, and the Agent's cottage, once a picture of life and beauty, its once beautiful garden plot covered with rank weeds'.

Local information relates how a great lake of water was met in Lathkill Dale Mine, all tools had to be abandoned, and the miners made a hasty retreat. A letter written in 1884 by James Bacon, a working lead miner, states that at the last working day at Lathkill Dale Mine 24 loads of ore were made ready for the market (62). The small scale working of Lathkill Dale Mine after 1851 will be dealt with later.

2. Mandale Mine 1839-51

A new Company was formed in December 1839, by John Wakefield and John Alsop jnr. John Alsop, snr, died in 1836 and the family mining and smelting interests were carried on by his son, also John. The former, acting as agent to the Mandale Mining Company, was put into possession of the Mandale and Pasture Rakes, and several smaller veins and scrins associated with them (56). An ambitious Prospectus was issued about 1839-40 (57). The failure of Mandale Sough

Above and below: Two views showing the beginning of the leat which took water to the 'great wheel' at Lathkill Dale Mine and to the Mandale Engine Shaft water wheel.

5 Power to Drain the Mines

was acknowledged, as the old men's workings had virtually reached that depth before the sough was driven. Additionally, the old miners had laboured only with the use of primitive hand pumps, but

> 'the present Company propose to erect a steam engine of sufficient power to drain the mine sixty yards below the present level, and also erect a water wheel ... by surplus water from the reservoir of a neighbouring mine ... The wheel will be enabled to lift the water from a depth of thirty yards and thereby save the expense of working the steam engine...'

The 'neighbouring mine' was of course Lathkill Dale Mine, which itself was rapidly nearing closure. John Wakefield had already subscribed for 200 shares in the new company, each shareholder being asked for a £2 deposit per share.

Throughout the reckoning books are references to an 'Old Level' and this is also mentioned in an Estate Book:

> 'May, 1842. Examined the Mandale Mine "old man's level", the "1800 level", and the new "middle" and "deep levels". The old man's level got the ore upwards to the surface and must have been rich. The 1800 level was driven an average $1^{1}/_{2}$ yards below the old man's level and could not therefore answer' (9).

The 'old man's level probably refers to the eighteenth century Pasture Rake Sough, the '1800 level' must be Mandale Sough and the (then) new 'middle and deep levels' equate to two levels driven up Mandale Rake beneath the sough and thus beneath the river level.

Building the Aqueduct

In 1840 the Mandale aqueduct was constructed. It carried water from the leat on the south side of the River Lathkill, by means of wooden troughs or launders laid across the tops of the six stone pillars, to a short leat on the north bank and so to the wheel pit behind the later built Cornish engine house. The water wheel was of approximately 35 feet diameter, and pumped from the Bottom or Deep Level, 90 feet below the level of the Lathkill. The pumps were 14 inches in diameter and the water was discharged into Mandale Sough. The wheel was definitely at work by March 1841 (58).

A report made to the mine proprietors in May 1842 gives a clear

5 Power to Drain the Mines

Remains of the aqueduct from the leat on the North side of the valley.

5 Power to Drain the Mines

description of the state of the mine at that time. The sough had been cleared out for a distance of 636 yards from the foot of the Inclined Plane and was open for 200 yards beyond. It had been found to be only about 3 feet high and 1 foot 6 inches wide and had therefore to be enlarged for its entire length. The vein was extremely rotten and 368 yards had to be laundered to prevent the water from sinking into the lower levels. There were two new levels below the sough, one at

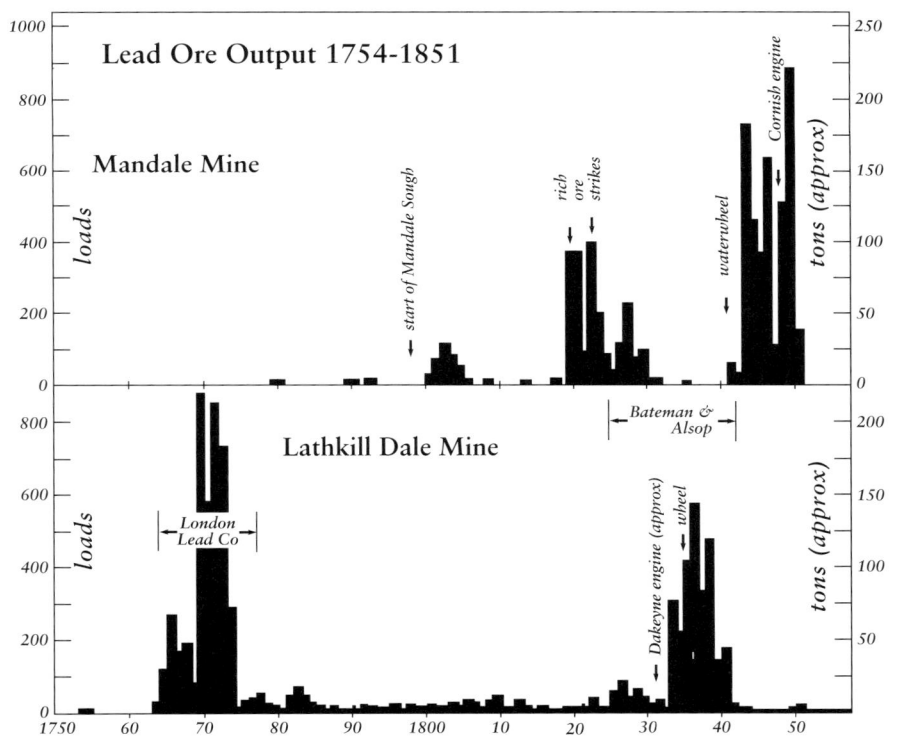

Lead Ore Output 1754-1851

Mandale Mine

Lathkill Dale Mine

The figures do not include any of the smaller mines in the Dale. The very low output during most of the period, with the occasional high peak when a rich strike was made, is typical of the Derbyshire lead mines. The effects of the installation of the water wheels and the steam engine is clearly demonstrated. Often different sources of production figures do not tally, for instance during the rich strikes of 1820-21, 722 loads of ore were produced at Mandale Mine yet none is recorded in the Barmaster's books! Over the 100 years, the two mines totalled 14,100 loads of concentrated ore, less than 3,700 tons.

9 fathoms, or Middle Level, and the 15 fathom, or Bottom Level. At the time of the report the driving of the Middle Level had been suspended 128 yards from the Engine Shaft. The Bottom Level was poor in ore but they were within 70 yards of where rich ore had been left in water in the upper levels (58).

Two payments totalling £175 made between January and June 1842 to the Butterley Company, Ripley, were thought originally to have been for parts of the water wheel, (59) but this now appears very doubtful.

Mention has to be made of the datestone originally incorporated in one of the stone pillars of the Aqueduct. This now bears the date '1810', a date which in the past has led to considerable confusion. Close inspection demonstrates that it has been defaced at some time, the date it originally bore being 1840, the year the pillars were built. This date is also substantiated in an old Estate Book:

> 'The Mandale pillars for carrying the launders across the Lathkill, appear by a stone fixed in them to be built in 1840'.

Strangely enough this old book originally added to the confusion. The handwriting in parts is very poor and cramped, and 1840 was misread as 1849 (9).

The rich ore hoped for in the 1842 report seems to have been located, because in 1843 the Company measured 719 loads and 3 dishes of ore. Production continued at a fairly high level until 1849, with the exception of 1847 when only 102 loads and 8 dishes were measured (60).

The Cornish Engine

By this time pumping by means of the water wheel was inadequate and a Cornish Engine was ordered. The engine house and boiler house were built in 1847 from stone quarried a little higher up the Dale (9). The engine was built in 1848 by Messrs. Graham and Co. of the Milton Ironworks, Elsecar, near Barnsley. It was of 150 horse-power, constructed on the three-valved Cornish expansive principle. The cylinder was 65 inches diameter, with a 9 feet stroke, 8 feet $6^{1}/_{2}$ inches in the pump. There were two tubular boilers, 25 feet long and 6 feet 6 inches diameter (61).

There followed two good years of ore production, but the high capital cost of the plant and operating costs forced the Company to wind up its activities in 1851. The last payment to Lord Melbourne for the water lease was made in November 1851, and the whole of the mining

5 Power to Drain the Mines

The aqueduct from the leat on the north bank.

property, including the steam engine and boilers, water wheel and numerous miscellaneous wagons, wrought iron rails, etc., were offered for sale by auction on the 28th January, 1852. Unfortunately no reports have come to light after the sale, so the fate of the steam engine and water wheel is not known. The late Mr. Rowland of Conksbury said that he understood from his father that 'a pump from the Mandale Mine went to Calver Sough Mine.' Mr. Rowland also said that his father, who was born in 1841 and could therefore remember Mandale Mine working, had told him that the water emerged from the pump delivery pipes under such pressure that it looked like new milk. Another story handed down from the old residents of Over Haddon is that it took forty horses to pull the beam of the Cornish engine up the hill from Lathkill Dale bottom to the village.

The Estate Book states that on giving up the mine the Mandale Company removed the machinery, boilers and other fittings, but were told in December 1852 that they had no authority under the *High Peak Mineral Customs and Mineral Courts Act* to demolish the buildings.

One reference states that the Mandale Mining Co. lost £36,000, but the period during which this loss accrued is not given.

Later Activities

Mandale Mine continued to be worked until 1867 by a large number of independent miners, including the old agent James Bateman. A little of the ore produced was of good quality, but by far the greater proportion was of very inferior grade. No doubt it would be mixed with a large amount of iron oxide.

Lathkill Dale Vein, and to a lesser degree Smallpenny Vein and Sideway Vein, were worked in similar fashion during this same period. For example, in 1858 John Millington measured 114 loads 2 dishes of ore at Smallpenny Vein but at an average price of only 1s. per load. At Mandale in 1865, Anthony Brocklehurst obtained 18 loads 7 dishes of better quality ore at £3 9s. $4^1/_2$d. per load (60).

The Decline: Lathkill Dale Mine and Vein after 1851: Gank Hole and Lathkill Dale Level

Although working on a large scale had ceased at Lathkill Dale Mine in 1842 and the great wheel and other machinery auctioned in 1847, lead ore continued to be raised for a good many years afterwards. A small amount, 24 dishes, was measured in the years 1843 and

1844, after which there was silence until 1850. Rather surprisingly, in that year 55 dishes were measured, but still in the name of the old Company and a further 92 dishes measured the following year. This was the last recorded working at the mine by Alsop and Co., but concurrently the agent, James Bateman, commenced working for himself. He did not work regularly at the one mine, but appears to have moved around working at Sideway Vein and more particularly Mandale Mine.

Several other miners were also extracting lead from Lathkill Dale Mine during these years, but the yield was never large and furthermore a significant amount of the ore raised was of inferior quality. The last ore measured was in 1865 when William Taylor obtained 6 loads 8 dishes.

Gank Hole Vein was the scene of the last large scale mining adventure in Lathkill Dale. The mine was worked intermittently for limonite (iron hydroxide), used as a source of pigment, and samples had been assayed in 1855 to determine whether the metallic iron content was sufficient to warrant its use as an iron ore. Unfortunately, it proved to be too inconsistent.

The Gank Hole and Greensward Venture

A rather grandiose scheme was mooted in 1884 when it was proposed to drive a level from the Lathkill, along Gank Hole Vein, into Mycross Vein and finally into the Great Greensward Mine. A company titling itself the 'Mycross Mines and Lathkilldale Level' was formed and work began. The level was to serve the dual purpose of a wagon gate and drainage level. To begin with they were raising 3 to 5 tons of ochre (limonite) per fathom, the driving costing £1 a fathom. Adopting an extraordinary sense of inflated optimism, it was believed that 40,000 tons of lead ore would be got in the Greensward Mine alone, and it was estimated that the whole venture would bring a return of £2$\frac{1}{2}$ million (62).

The principal shareholder was James Farnsworth, who at the time the prospectus was issued lived at Longsight, Manchester. He became involved in various disputes regarding ownership of the old Agent's House in Lathkill Dale and was also sued in the Small Barmote Court for non-payment of wages by some of his miners. The level was a failure and was only driven 230 yards from Lathkill Dale. An attempted

Above and below: Gank Hole Mine, circa 1884. The building with the chimney was between the path and the river. Note the water pipe (above) bringing water across the river from the leet.

The lift pump chamber in Greensward Mine at 55 fathoms depth. On the left can be seen the surviving pump rising main and the upper valve box.

resuscitation of the venture was made in 1889, but so far as is known nothing materialised. Mining was still going on in 1891, because in that year Tom H. Brown, the manager of the Greensward Mining Company requested permission from the Estate to erect a small water wheel or turbine on the river to provide power for Lathkill Dale Level. No official reply from the Estate can be traced (64).

Greensward Mine itself is perhaps a little too far away from Lathkill Dale to be discussed in detail here, but it might be mentioned that the

5 Power to Drain the Mines

vein, variously spelled Greensa, Greensaw, Greensward, Greensor etc., was working in the sixteenth century. References occur throughout the eighteenth and nineteenth centuries, terminating with an attempt to work the mine to greater depth by means of a horizontal steam engine. This engine was utilised for the dual purpose of winding and pumping: no doubt the pump rods in the shaft would be connected to the piston by means of a sweep rod, in similar fashion to the Mitchell engine at the Peak Forest Mine. The pumps and associated pitwork still remain in the shaft (see Buckley and Howard in Bulletin Peak District Mines Historical Society, Vol.12, No.6, 1995, for a full account). The engine was in the course of erection c. 1884-1887, the shaft being sunk to 60 fathoms (63).

The late Mr. Charles Millington, an old Monyash lead miner, remembered the engine working. He said that the engine house was built of brick and described the engine as a 'stand engine.' He remembered the mine ceasing work, his father having been employed there. The engine and other materials were auctioned, but he did not know what became of them. According to Bulmer's *Directory of Derbyshire*, 1896, large deposits of ochre were being mined at Greensward, some from narrow fissures and some from huge caverns. He noted that the supply seemed 'inexhaustable.'

The failure of the Lathkill Dale Level, coupled with the gradual closure of the other mines in the district meant the end of an industry spanning at least seven centuries. From time to time speculators have toyed with the idea of re-opening the mines. In recent years much 'hillocking' has taken place, particularly along Mandale Rake between the Top Engine Shaft and the Pasture Shaft and that part of the vein has been obliterated.

6 Features Visible Today

The Underground Remains

Much remains in Lathkill Dale in the way of accessible underground workings. There are many shafts and levels, many in a dangerous condition. It must be stressed that the following descriptions in no way indicate that any public access is encouraged into these ancient workings. Some mines are situated on land now within the National Nature Reserve and a permit must be obtained before the surface can be walked other than on the public footpaths. Permit arrangements exist for mine exploration with Derbyshire Caving Association and Peak District Mines Historical Society Ltd. Exploration should on no account be attempted other than through membership of these organisations and only then with expert guidance.

Mandale Mine

A detailed description of the underground features of Mandale Mine in Lathkill Dale has already been published, (65) so therefore only a brief sketch is included here, mainly noticing features of historical or geological relevance.

There are at present two modes of entry into the mine, either by Mandale Sough or by the Inclined Plane. The sough is open from its tail to a run-in in a small scrin vein parallel and very close to Pasture Rake, a distance of a little over 140 yards. For much of this length it is lined with dressed limestone. A nicely lined chamber on the east side of the Lodge Shaft is now partially filled with rubble, but it is thought that it once held the machinery associated with the pump rods of the water wheel. The walling and arching in this part of the sough dates from the 1840s at

Above: The former black marble mine now houses a colony of bats. **Below:** Virtually nothing remains of Carter's Mill.

6 Features Visible Today

which time it is known the Mandale Mining Company were obliged to re-drive a long length of the existing sough.

A shaft, shown on a section of the mine as the Engine Shaft, is situated on the bench of limestone immediately outside the entrance to the Inclined Plane. This shaft, now covered for safety reasons, communicates with the sough by means of a short, slabbed cross-cut. Below the sough the mine is flooded, but originally the shaft descended to the bottom level of the mine.

The Inclined Plane reveals very narrow old workings at several points in its roof, no doubt dating from at least the mid-eighteenth century, although some stopes are undoubtedly much older. The incline may have been built for hauling ore out of the mine. They were unusual in Derbyshire, but several survive and there are also three known in the North Staffordshire ore field. From the foot of the Inclined Plane, the sough ranges generally north westwardly, in part driven in solid limestone parallel to, and a few yards on the north-east side of the vein, and in part actually in the vein itself. The sough, or the greater part of it, between the tail and the Deep or Forefield Shaft, was enlarged during the 1840s. A descent of the Forefield Shaft, during recent years, revealed that the level beyond is very small so it does not appear that the re-driving was continued past this shaft.

The stopes off the sough cannot be dated with certainty, but they are interesting in demonstrating that parts of the vein were worked in a very large, red-brown, clayey brecciated deposit, almost developing into a pipe. A little way beyond the sough is driven in the vein, which here has changed character to a thin stringer of baryte with a little galena. Old workings dating from the eighteenth century are visible in the roof, and periodically the vein opens out slightly in the loose material. The limestone walls exhibit horizontal slickensiding, the rock itself being thinly bedded with chert bands clearly visible. This rock, the 'Blackstone' of the old lead miners, is the Lower Lathkill Limestone or Monsal Dale Limestones Dark Beds of modern geologists. A large fall of rock from a stope above occurs but this has been opened by explorers and a connection made with workings leading through to the Forefield Shaft, 770 yards from the tail in Lathkill Dale.

The Forefield or Deep Shaft was descended in the mid-1960s and 1973 and the following points are perhaps the most interesting. Upstream from the shaft the sough was explored for approximately 800 yards, but was found to be in poor condition. Most of it was driven in the vein and was arched or slabbed over, having a cross section of 4 feet high by 2 feet 6 inches in width. No evidence of working below the sough level was noticed. There may be from one to two feet of running water flowing along the sough which issued from a terminal run-in. Downstream from the shaft all the water flows through deep, flooded stopes before reaching the connection with the working usually explored from Lathkill Dale. A small dressing floor was found together with a few tools. Again part of the level is in the vein and the section is about 400 yards in length southeast of the Forefield Shaft. For more information see Ford and Worley, 1976, Bull. PDMHS, Vol. 6, No. 3, pp 141-3.

Lathkill Dale Vein and Sough

The main way of entry into the workings along this vein and into the sough, is via the large shaft, probably the Dakeyne pumping shaft, beneath Bateman's House. There are two large shafts side by side, the first is filled with rubble to within 12 feet of the surface. At this depth a short level, walled and arched with dressed limestone blocks, gives access to the second shaft, directly beneath the ruins of the agent's house. This shaft is 12 feet in diameter and descends directly to the Lathkilldale Sough. There is a good depth of rubble in the bottom of the shaft, where a large excavation was probably connected with the Dakeyne engine installation.

Lathkill Dale Vein can be seen in the shaft bottom as a two feet wide fault fissure filled with cream coloured calcite and some broken limestone, hading to the north at about twenty to thirty degrees. The limestone is thinly bedded, dark coloured and sharply turned upwards at the fault. It is the Lower Lathkill Limestone, probably the 'Blackstone' of Mandale Mine. Upstream the sough can be followed for some 260 feet in a fairly irregular fashion. At this distance a brick dam is reached, completely sealing the sough beyond. This brickwork was done in 1854, after the main closure of the mines, in an attempt to prevent the water in the river from sinking into the old workings.

Under normal weather conditions it is impossible to penetrate more than a few yards into the sough. However, in the prolonged drought of 1959 it was possible to follow the sough downstream for

6 Features Visible Today

plan

Lathkill Dale Mine

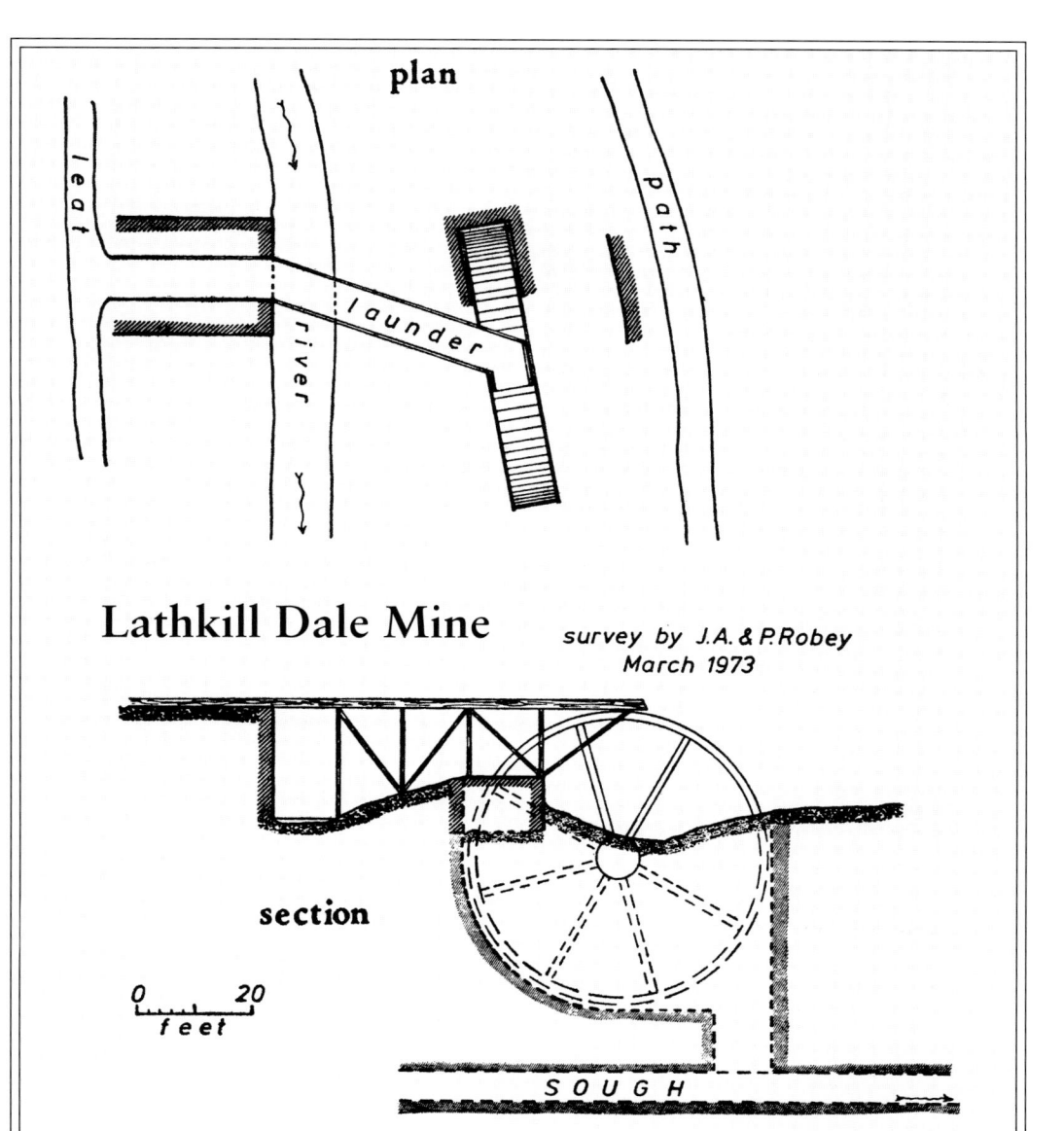

*survey by J.A. & P. Robey
March 1973*

section

0 — 20 feet

SOUGH

Survey and partial reconstruction of the water wheel pit at Lathkill Dale Mine. No attempt as been made to show the probable connecting rod and angle-bob beam for pumping water up to the sough.

a considerable distance. Unfortunately, no survey was made but it was estimated at the time that about 1500 feet of the level was explored. Much horizontal slickensiding was in evidence, and one rather unique feature is the fact that the whole of the level has been driven utilising the hade of the vein so that the passage leans for the entire length explored. Eventually the water became too deep for further exploration, there being only a few inches of air space (67). The sough was also bricked up in 1854 at a shaft just below the Sough Mill, and is the reason for the backing up of the water.

Miscellaneous Levels and Workings

A conspicuous level at SK 188658 is driven in thinly bedded Lower Lathkill Limestone (or the Monsal Dale Limestones Dark Beds) from the side of the river to intersect the Sideway Vein. One old map (68) marks the level as a sough, but this is somewhat doubtful. More probably it was driven as a haulage gate, but may well have served the dual purpose of haulage and drainage. The old stopes in Sideway Vein accessible via this level are very narrow and partially packed with deads. With extreme care it is possible to climb through these old workings and emerge much higher up the hill via some open cuts in the vein. Some of these workings are probably those excavated by Richard Glossop and partners during the 1750s and 1760s.

A small level at SK 183658 is driven along the westward continuation of Gank Hole Vein. Just over 190 feet in length it is quite narrow with some small, narrow stopes above. The vein contains much calcite with limonite, but the mine is of more interest in that it contains traces of the mineral rosasite, an uncommon copper and zinc mineral.

There are many smaller levels and shafts, and a large number of shafts still await exploration. Some workings have not been described as they are in an extremely dangerous condition and exploration is definitely discouraged.

The Surface Remains

Despite being relatively shallow, the lower workings of the Mandale Mine were only 90ft. below the valley floor, the mines were always troubled by large quantities of water that accumulated underground. This is hardly surprising when the close proximity of the River Lathkill and the cavernous nature of the

(cont'd on page 87)

Mandale Sough (below) and a shaft rising above it (above). Both views near the sough tail.

6 Features Visible Today

Near the entrance of the level on Sideway Vein with David Arveschoug.

6 Features Visible Today

limestone strata is considered. Moreover the Lathkill Dale Vein runs directly along the course of the river for much of its length so direct seepage must have been a constant problem. Thus much of the mining activity in the dale was directed towards keeping the workings dry and it is the physical remains of these efforts that are most prominent today.

Lathkill Dale Mine

The site of the Lathkill Dale Mine is marked by large overgrown hollows between the river and the pathway through the dale. These hollows, often filled with water in wet weather, are the collapsed remains of the water wheel pit and associated shafts. The point where the leat from Carter's Mill crossed the river is marked by a conspicuous dry-stone wall and part of the walling for the wheel-pit is clearly visible. The water from the leat would probably be carried to the top of the giant water wheel in wooden launders mounted on wooden trestles (Fig 9). A consideration of the head of water available to work the wheel with the waste water being discharged along Lathkilldale Sough shows that much of the water wheel would have been contained in a very deep pit with only a small portion showing above ground. Nevertheless the wheel would certainly have been an imposing sight when at work.

Without extensive excavations the archaeological remains are not sufficient to give any details of the mechanism which connected the water wheel cranks to the pump rods in the shaft. No doubt the drive would be via connecting rods and angle-bobs or beams to pump from below sough level, with much of the machinery in excavations underground.

Mandale Mine

At the Mandale Mine the combined use of a drainage sough, water wheel and steam engine to keep the mine free from water is clearly demonstrated at what is the best site of its kind in the Peak District. The pit which housed the 35ft. diameter water wheel can be clearly seen on the north-west side of the Lodge Shaft. The water supply came along the leat from the pool at Carter's Mill, over the aqueduct, through a short arched tunnel (probably to allow a trackway for materials from out of the mine to pass over the leat) and onto the wheel, probably via a wooden launder. As the wheel worked pumps 90ft. below the sough and the steam engine was intended to pump from twice that depth

(cont'd on page 90)

site plan

Survey and simplified reconstruction of the Cornish engine house and water wheel pit at the Mandale Mine. Surviving features are shown as bold lines while the reconstruction is shown as finer lines.

Mandale Mine

survey by J.A.Robey, J.Mathews & P.Robey
March 1973

suggested method of coupling waterwheel to pump rod

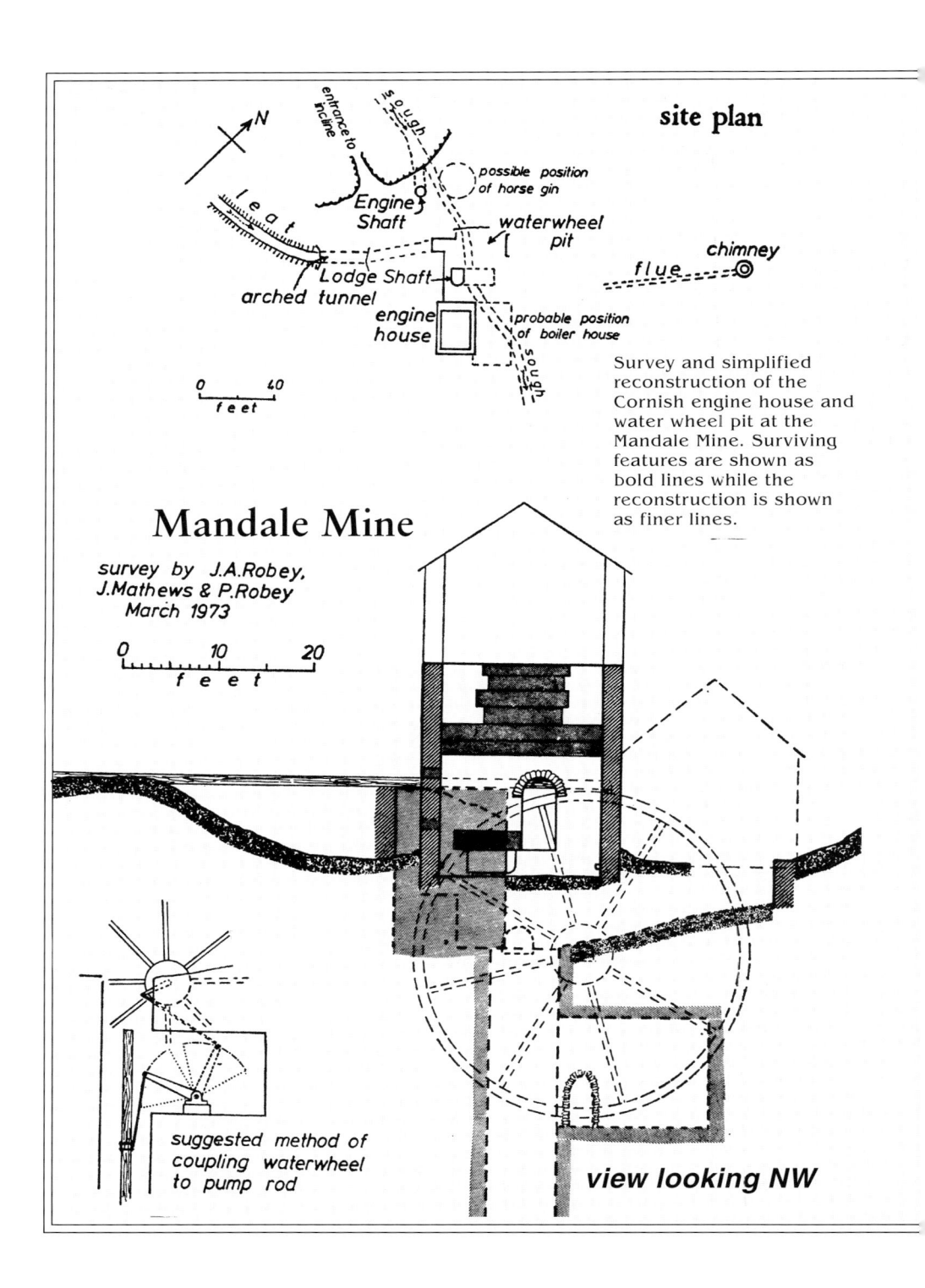

view looking NW

plan

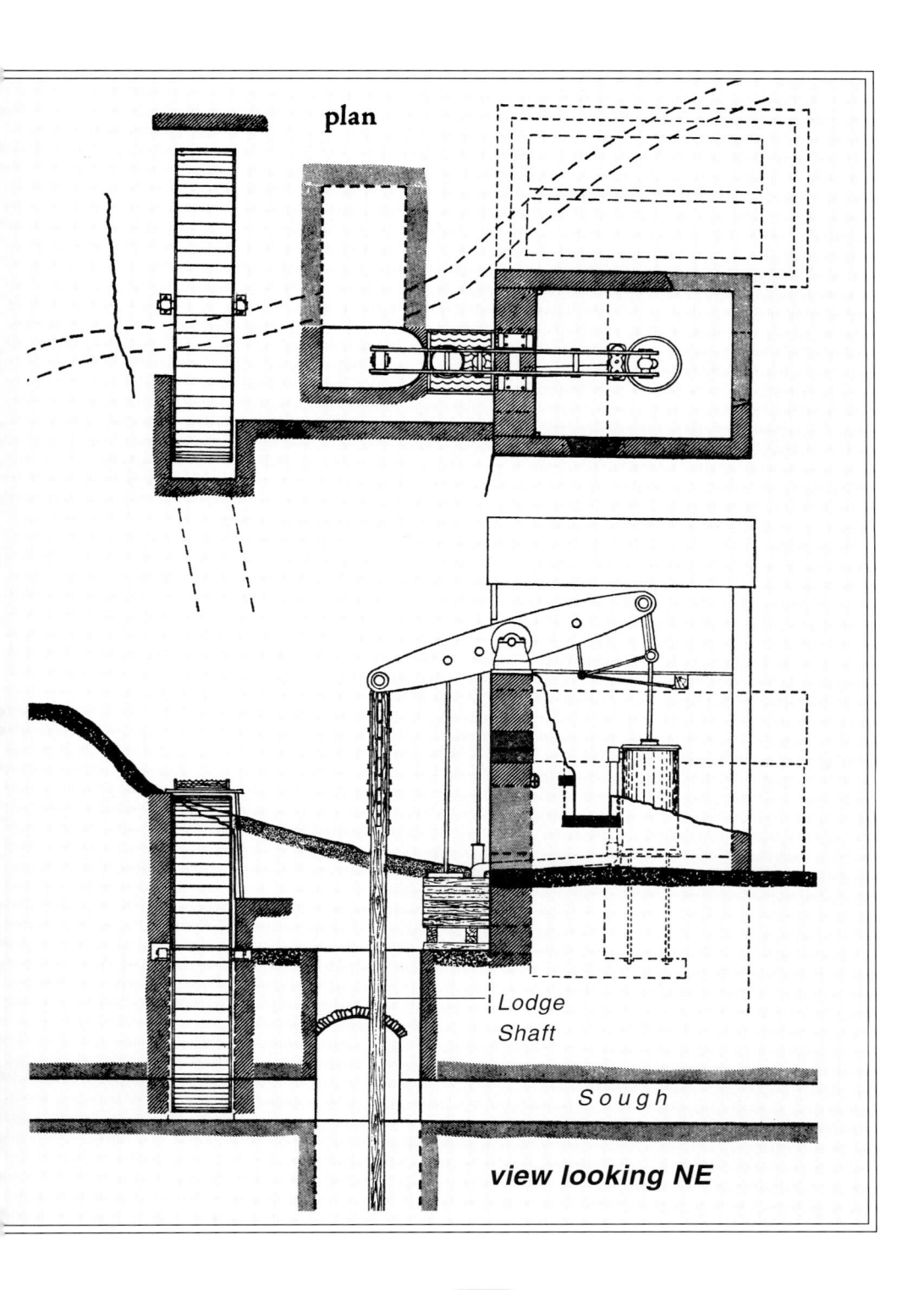

Lodge
Shaft

Sough

view looking NE

(although there is no evidence that it ever went lower than the 90 feet or Bottom Level) it is probable that separate pump rods were employed. It is likely that the reciprocating motion to operate the pump rods was transmitted from the wheel crank by means of an angle-bob with one part in the wheel pit and the other part in the underground chamber at the side of the shaft. Probably the axle of the angle-bob passed through a passage-way into the chamber, but as this is now nearly full of rubble this could be proved only after excavation. It is likely that the tail-race water from the wheel was returned to the sough via this passage.

Due to the splintery nature of the limestone used to build the Cornish engine house this has largely collapsed leaving only the very substantial bob-wall upon which the cast iron beam rocked. The layout of the engine within the building can be seen from the reconstruction (Fig. 10). The massive foundations and walls needed to support the steam cylinder and beam can be appreciated from the diagram.

The chamber at the side of the shaft at sough level has often been thought to have contained a balance-bob to counteract the great weight of the steam engine's pump rods, which would be made of pine up to l8in. square and descend to the lowest depths of the shaft. The plunger pumps which would be used relied on the weight of these rods to force the water to the surface, the role of the engine being simply to lift the rods every cycle. Only when the weight to be lifted was too great for the engine were balance-bobs employed. With the Mandale engine having relatively short pump rods it is unlikely that any balancing effort would have been necessary. A more probable use of this chamber seems to be for coupling the water wheel to its pump rods and to allow the waste water from the wheel to flow into the sough as described above.

The base of a round limestone chimney, nearly 9ft. in diameter, can be found on the hillside to the north-east of the engine house. About 90ft. of flue (4ft. 6in. high and 2ft. 10in. wide internally) can be traced from the chimney to within 70ft. of the engine house. The conclusion is that the now completely ruined boiler house was built onto this side of the engine house. There is a flattish area here which was probably the location of the boiler house and in the confines of the site this is the only practical position for it. Although the Mandale engine house is the most prominent feature of the site it was one of the shortest lived for it was in use only for about five years.

Chronological Table

1284	Mandale Mine at work
1288	Mandale Mine mentioned in the 'Quo Warranto'
1292	Leicester Abbey owned mines in 'Medoplek' (Meadow Place Grange)
1495	Mining taking place on both sides of River Lathkill
1585	The 'best ore in the Peak' was being mined at Over Haddon Field Rake
1615	Some workings at 'Mem Dale,' presumably Mandale, were 300 feet deep
1632	Mining in Over Haddon Pasture
1635	Enquiry regarding the possibility of erecting a lead smelting mill on the river
1640	Mandale Rake ore worth 16/-per load
1649	Earl of Devonshire agreed to eleven Articles being drawn up to govern lead mining in Meadow Place Grange
1665-66	Mandale Rake at work within Bakewell Liberty
1677	700 loads of ore illegally removed from two meers in Mandale Rake. The mine was 90-120 feet in depth
1694	Great Barmoot Court held at Upper (or Over) Haddon
c. 1727	Two water wheels driven by the River Lathkill, situated on Lathkill Dale Vein
1730	Over Haddon Suff (sough) mentioned. The first drainage level to be recorded in the valley
1743	Lathkilldale Sough probably commenced
1744-49	Bristol adventurers and merchants working mines in the Dale
1750	Lathkilldale Sough had reached Meadow Place Grange

Chronological Table

The bob-wall of the Mandale Engine House showing the inner side

Chronological Table

1761	London Lead Company acquired the title to Smallpenny Vein in the higher reaches of the Dale
1764	London Lead Company acquired the title to Lathkill Dale Vein. Lathkilldale Sough had probably reached the site of the later shafts at Bateman's House
c. 1769	A hydraulic engine, or Water Pressure Engine, invented by William Westgarth possibly installed at Lathkilldale Mine
1770-73	High output of lead ore from the Lathkill Dale mines owned by the London Lead Company
1776	After a rapid decline in production, the London Lead Company decided to sell their holdings in the mines in Lathkill Dale
1779	The Hill Carr Sough Company acquired title to the 'Lathgilldale' mines, comprising of some 36 separate veins with a combined total length of over twenty miles
1782	Two Over Haddon miners in possession of Lathkill Dale Vein. Lathkilldale Sough by this date was just over a mile in length
1787	Beresford's Level (Ricklow Mine) commenced, supposedly with the intention of draining Magpie Mine
1797 or 1798	Mandale Sough commenced
1807	Mandale Sough 700 yards long
1820	The first of two large ore strikes at Mandale Mine. In 26 weeks £1,155 profit was made
1823	The second large ore strike. Profit of £585 was made
1825	Thomas Bateman and John Alsop, senior, bought Lathkill Dale Vein from the Taylor family for £30
1827	Bateman and Alsop took a 21 year lease from Lord Melbourne at £25 per annum for the privilege of using water from the river for driving water wheels and permission to erect other engines for pumping water
1831	A hydraulic engine, patented by the Dakeyne brothers of Darley Dale, was in course of erection, very probably at Lathkill Dale Mine

Chronological Table

1836	Erection of the 52 feet diameter water wheel at Lathkill Dale Mine, built by Benjamin Smith & Co. at the Duckmanton Ironworks, near Chesterfield
1836	George Hollis of Youlgreave, miner, killed by a fall of stone in Mandale Mine
1836	James Bateman became agent to the Lathkill Dale Mine
1838	All work at Mandale Mine under the old company virtually ceased
1839	New company formed to work the Mandale Mine
1840	Aqueduct built and water wheel of 35 feet diameter erected at the Mandale Mine
1842	Lathkill Dale Mine ceased large scale working
1843-50	High output of lead ore from Mandale Mine
1847	The large water wheel and the rest of the property offered for sale at Lathkill Dale Mine
1847	The engine house and boiler house built at Mandale Mine
1848	Cornish engine, 150 horse power, 65in.-cylinder and 9ft.-stroke, built by Messrs. Graham at the Milton Ironworks, Elsecar, near Barnsley, for the Mandale Mine
1851	Mandale Mine ceased work after a reputed loss of £36,000
1852	Plant, machinery and mining property auctioned at the Mandale Mine
1854	The old Lathkilldale Sough bricked up in two places, to prevent the river water from sinking into the old workings
1865	Last recorded measure of lead ore at the Lathkill Dale Mine
1867	Last recorded measure of lead ore at the Mandale Mine
1884	Lathkill Dale Level being driven up the Gank Hole Vein
c. 1884-87	Steam pumping engine in course of erection at the Great Greensward Mine

1890s to early this century. Sporadic attempts to work ochre with lead as a by product at Greensward Mine and at Lathkill Dale Level

Mining Development in the Dale

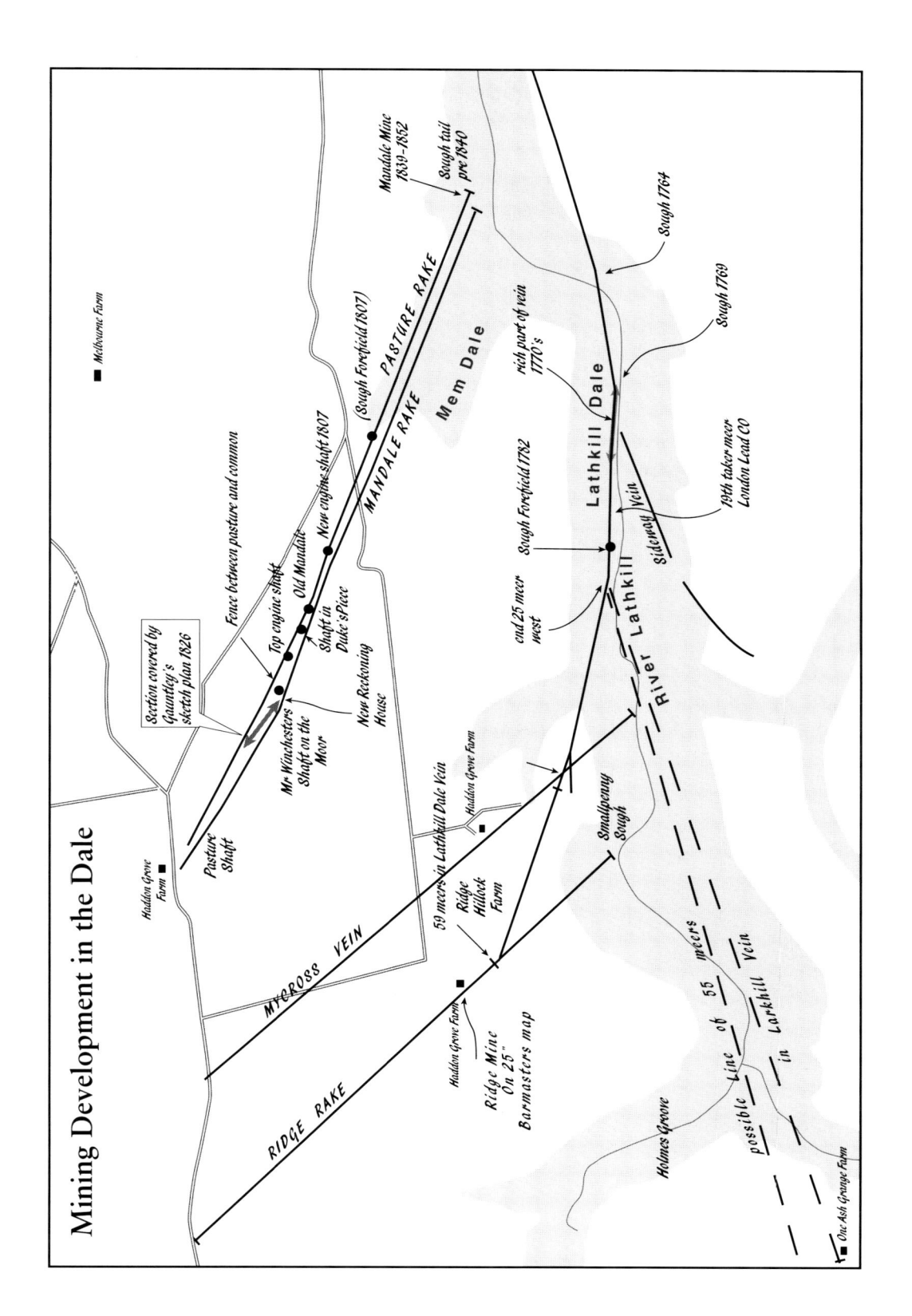

Glossary of Mining Terms

Adit
A horizontal tunnel or gallery, sometimes used as a haulage road, sometimes as a drainage level (see also cartgate and sough).

Adventurers
Shareholders or financiers in mining ventures, often wealthy landowners or industrialists.

Bargain
An agreement made between the mine owners (generally by their agent) and the miners to undertake a specified amount of work at a fixed payment. (See also cope-bargain).

Barmaster
A Crown official, responsible for the measuring of meers along a vein and the relevant entries of the mining title in his book of records, the measuring of lead ore before sale, the convening of the Barmote Court and the general administration of mining law.

Barmote (Barmoot) Court
Presided over by a Steward and the Barmaster, the Jury is comprised of twelve (formerly twenty-four) miners or persons connected with mining. Described as being of great antiquity in 1288, some of the laws may date from the Roman occupation. The Court is still in existence, that for the Soke and Wapentake of Wirksworth is held yearly in the Moot Hall there, others being held for other districts or Liberties in the lead-mining field, for example at Chatsworth House and occasionally at Haddon Hall and elsewhere.

Barytes or Barite
The mineral barium sulphate known to the miners as cawk or heavy spar $BaSO_4$.

Belland
Very fine particles of lead ore resulting from the crushing process prior to smelting. Cattle may be poisoned by it and are then said to be 'bellanded.'

Bing
Large pieces of ore as raised from the mine and requiring little cleaning.

Black Jack
Sphalerite or zinc sulphide ZnS.

Blackstone
The miner's term for either a dark bituminous limestone, as in the Monyash-Over Haddon areas, or for lava or toadstone as in the Bonsall-Matlock-Wensley areas.

Glossary of Mining Terms

Blue John
An ornamental form of the mineral fluorite, found only at Treak Cliff, Castleton.

Bole, Bole hill
A primitive smelting site, consisting of a circle of stones in which alternate layers of wood and ore were placed. The blast hole always faced the prevailing wind, hence boles were usually sited on the western edges of the hills. Due to the poisonous nature of the fumes, the boles were placed in very remote locations.

Bouse
Lead ore as drawn from the mine.

Bucker
A flat headed hammer used in breaking down the lead ore on the dressing floor.

Buddles, buddling
The process of washing fine waste by a stream of water through a series of troughs, in order to extract the fine lead ore discarded by former miners.

Calcite
The mineral calcium carbonate $CaCO_3$.

Caulk or cawk
The miners' term for barytes.

Calamine
Zinc carbonate, known to miners as 'bone ore.' Once used extensively in the manufacture of brass.

Cartgate
A horizontal level in a mine, usually the main haulage road. A good example is at the Odin Mine, Castleton.

Chert
A rock similar to flint, and with the same chemical composition, found as layers and nodules in limestone. Due to its hardness great trouble was often caused in driving through it.

Clay-wayboard
Thin layers of clay between limestone beds, resulting either from the decomposition of lava, or formed from fine dust thrown out by volcanic explosions.

Coe
A small cabin at the mine, used by miners for changing their dress, storing their tools and ore. The climbing shaft was often situated in the coe.

Cope-bargain
An agreement for miners to extract, and prepare for sale at their own cost, a specified quantity of lead ore at a fixed price per load.

Above: The bridge which carried the cart road up to the inclined entrance to the mine and the water wheel. The latter was situated at the far end of the bridge.
Left: Remains of the water wheel pit at the rear of the Mandale Engine House. The Mine was kept clear by water power whenever possible, using the expensive engine only when necessary.

Cope
A royalty paid so that miners may sell their ore to whom they please, the Crown or its lessees having first right of purchase.

Coper
One who mines ore on a cope-bargain.

Cross-cut
A level driven in barren rock branching out of a vein or a sough.

Cupola
A reverberatory smelting furnace.

Deads
Waste rock, stored under ground whenever possible in cavities from which the ore has been extracted. When drawn to the surface they are thrown on the hillocks.

Dial
The miners compass.

Dish
Nearly all lead ore was measured by volume. In the Low Peak a dish of ore contained 14 Winchester pints, in the High Peak 15 pints. The weight of lead ore per dish varied according to quality, but averaged about 65 lbs. Nine dishes made one load. Four loads made a ton.

Fang
Pipe for conveying fresh air into the workings.

Glossary of Mining Terms

Fathom
The unit of length and depth in mines; equal to 6ft.

Fault
A structural weakness in the earth's crust resulting in movement of the strata. Rakes or veins are usually mineralised faults.

Firing
The practice of lighting fires against the rock face in order to break it down. The practice was discontinued in Derbyshire about 1720-1730. Fires could only be lit underground between 4 pm and 8 am due to the danger of suffocation from smoke.

Flat
A mineral deposit in which the ore and associated minerals are enclosed between the bedding planes of the country rock.

Fluorspar
The mineral calcium fluoride CaF_2.

Forefield
The working face in a level or stope.

Founder (-stake, -shaft, -meer)
The original place of breaking ground in a vein. All meers were measured from the founder. When a new vein was discovered two founder meers were allocated to the miners. If an old vein was reworked only one founder meer was given. The founder shaft was always situated in the first founder meer. (See also Lord's meer and taker meers).

Freeing
Each meer of ground along a vein, whether in an old vein or a new, had first to be freed by the handing of a dish of ore to the Barmaster, before any mining could commence in that meer.

Galena
The commonest form of lead ore found in Derbyshire. Composed of lead sulphide, it is a metallic grey mineral and can be found in most old mine dumps and hillocks.

Gangue
The minerals accompanying the lead ore in the veins. In Derbyshire the three commonest gangue minerals are calcite, fluorspar and barytes, although others such as iron minerals occur less frequently.

Goethite
Iron Hydroxide $Fe(OH)_2$.

Gin
A primitive winding apparatus, consisting of a horizontal wooden drum revolving around a vertical centre post. Ropes wound around the drum passed

over pulleys at the head of the shaft. The buckets or kibbles containing ore or waste rock were attached to the end of the rope by means of a swivel hook or clevis. The whole piece of machinery was worked by a horse or horses walking in a circle (the gin-circle or gin-race). A contraction of engine.

Ginging
The stone lining at the top of a shaft which continued down until firm rock was reached. Below this horizon the shafts were not usually lined.

Grand Jury
The members of the Jury of the Barmote Court, formerly 24 in number, but reduced to 12 after the passing of the two Acts of Parliament relating to Derbyshire lead mining in 1851 and 1852. Also called the 'Body of the Mine' or 'the twentyfour men'.

Grove or groove
An alternative name for a mine.

Hade
The lean or inclination of a vein from the vertical.

Hillocks
The tips of waste stone and gangue mineral discarded after the lead ore had been extracted.

Hoppit or dish hoppit
The wooden dish, identical with the standard dish, in which the ore was measured at the mine.

Jaggers
Teams of ponies employed to carry ore from the mines to the smelting mills. There are still several 'Jagger's Lanes' in Derbyshire.

Kibble
A bucket used for drawing ore and rock up the shaft.

Knock-stone
A large flat piece of stone or a metal plate, used for crushing the ore and gangue when drawn from the mine.

Launders
Wooden troughs used for carrying water, both above and below ground.

Liberty.
An area, often identical with the parish, within the mining field. Either part of the Queensfield, or a private liberty with the mineral rights belonging to a private individual and not within the jurisdiction of the Queensfield Barmote Courts. For example, Nether Haddon is a private liberty belonging to the Duke of Rutland.

Glossary of Mining Terms

Limonite
An impure form of iron hydroxide.

Load
Nine dishes of dressed lead ore, weighing about a quarter of a ton.

Lot
The duty payable by the miners to the owners or lessees of the mineral duties. Normally taken as 1/13th of the dressed ore, although it was varied from time to time to suit prevailing economic conditions.

Lord's meer
The third meer in a new vein, belonging to the owners or lessees of the mineral rights. A value was placed upon it by the Grand Jury, the miners then had the right to buy the meer outright, or they could cut a passage through it in order to get access to their taker meers, but they were not allowed to sell any ore in so doing. Sometimes two half meers were set out at each end of the founder meers.

Meer
A measure of length along a vein, irrespective of the width or depth. The meer varies from 32 yards in length in the High Peak to 29 yards in the Low Peak. The private liberties have meers of varying length, some of only 28 yards length.

Nicking
The process of claiming a mine. Formerly the Barmaster visited the mine once a week and if no work had been done, he put a nick in the spindle of the stow or winding windlass. This process was repeated on three occasions, if no work had been carried out by the owner then the mine was given away, without further notice, to the new claimant.

Ochre
Iron oxide, found in considerable quantities in certain veins and formerly used as a pigment. A general term for limonitic clays.

Old man (t'owd man)
Previous generations of miners. Also used to describe former mine workings.

Pig
A cast ingot of smelted lead.

Pipe
A type of mineral deposit characterised by its cavernous nature. Many originated as infill into ancient cave systems, or as whole or partial replacement of the surrounding country rock.

Queensfield or Kingsfield
The area covered by the jurisdiction of the High Peak and Low Peak Barmote Courts. The mineral duties of lot

and cope are owned by the Duchy of Lancaster and are at times leased to various individuals.

Rake, rake-vein
The commonest type of mineral deposit in Derbyshire. Formed either by the infilling of minerals into a joint, or more frequently a fault fissure. Some rakes are several miles in length and were worked to considerable depths. See also scrin and vein.

Rider
An isolated mass or piece of country rock within a vein or rake, tending to split the vein into two parts. Also called a horse, stalch or cranch.

Scrin
A small rake or vein, perhaps only a few inches in width, and of no great length.

Shack (or self-open)
A natural cavity into which mine water was sometimes diverted, or sometimes used to take waste rock.

Shaft
The vertical entrance to a mine. Small shafts were used for climbing and had timbers across for hand and footholds; alternatively holes were cut into the ginging and/or solid rock. Shafts were also sunk purely for ventilation, particularly along the line of a sough. Larger and deeper shafts were used for haulage and mechanical pumping (engine shafts).

Slickensides
The grooves cut into the vein walls by movement of the strata along a fault plane. The grooves can be highly polished and one variety has explosive properties.

Sole
The floor of a sough or a working level in the vein. The deepest part of the mine.

Sough, suff
A horizontal tunnel or level driven to drain water from a mine into a brook or river, or occasionally into an underground shack.

Standard Dish
A brass dish made during the reign of Henry VIII and still kept in the Moot Hall, Wirksworth. Dishes used at the mines were gauged against the standard dish.

Stemple
Pieces of wood used either as ladder staves as in a climbing shaft, or as rock supports on which to pack deads.

Steward
A solicitor who presides over the Barmote Courts.

Glossary of Mining Terms

Stowes, stoces, stowces
A wooden windlass used for hauling kibbles up the shaft, or used to denote possession of ground along a vein, each meer being delineated by a pair of small possession stowes.

Stopes
The working places in a mine from which the ore is extracted.

Tail
The entrance to a sough.

Taker meer
The remaining meers along a vein after the founder meer(s) and Lord's meer. Each had to be freed with a dish of ore before mining could take place in that meer.

Tithe
A proportion of ore claimed by the Church, nominally one tenth. Within some liberties miners refused to pay tithe and in others there was vigorous opposition by the miners who felt it was an unjust duty.

Toadstone
The miners name for any volcanic or igneous rock. Also known as channel, cat-dirt, blackstone or clay in different parts of the mining field.

Twenty-four
The former Grand Jury of the Barmote Court, now only twelve in number.

Vein
A vein or near vertical fissure, with well defined walls of country rock containing lead and other minerals. (See also rake and scrin).

Water-gate
Alternative name for a sough.

Whimsey
A steam engine used for raising minerals and materials up the engine shaft to the surface. Very few were used in the Derbyshire lead mines.

Winze
An underground shaft connecting one level with another.

Bibliography

Whilst the utmost care has been taken to ensure accuracy, it has clearly not been possible to cite every reference. Research over a number of years has enabled much background information to be filled in; particularly valuable in this context being the Bagshawe Collection and Oakes Deeds in Sheffield Archives, the BrookeTaylor Collection in the Derbyshire Record Office, the Devonshire Collection at Chatsworth House, and also the Public record Office, Kew.

The following are recommended for much general information on Derbyshire lead mining:

Ford, T. D., and Rieuwerts, J. H., editors, *Lead Mining in the Peak District*, 2000 (Landmark Publishing, Ashbourne).

Kirkham, N., *Derbyshire Lead Mining through the Centuries*, 1968 (Barton, Truro).

Raistrick, A., and Jennings, B., *A History Of Lead Mining in the Pennines*, 1965 (Longmans).

Rieuwerts, J.H., *Glossary of Derbyshire Lead Mining Terms*, 1998 (Peak District Mines Historical Society, Matlock Bath).

The *Bulletins* of the Peak District Mines Historical Society , now known as *Mining History*, cover a wide range of mining topics, while the many and widespread articles of Miss Nellie Kirkham are invaluable in any historical research.

References

Abbreviations

WHC. Wager Holmes Collection, Sheffield Archives, formerly Sheffield City Libraries.

Bag Coll. Bagshawe Collection, Sheffield Archives.

OD. Oakes Deeds, Sheffield Archives.

Dev Coll. Mining papers in the Devonshire Collections, Chatsworth House.

Bar Coll. Books, plans and other documents formerly in the possession of the Barmaster, but now housed at Chatsworth.

Br-T Coll. Brooke-Taylor Collection, Derbyshire Record Office, Matlock.

Woll Coll. Wolley manuscripts. Volumes relating to Derbyshire lead mining. Additional manuscripts, British Museum. Microfilm in Derbyshire County Library.

Bull PDMHS. Bulletin of the Peak District Mines Historical Society.

1. WHC., No.36, Letters from John Wakefield to William Wager.
2. Bag Coll., No.444, Barmasters Book of ore measurement and entries, Taddington, Upper Haddon and other Liberties, 1770-1820.
3. Farey, J., *The Agriculture and Minerals of Derbyshire*, 3 vols, 1811-1817.
4. Shirley, J., 'The Carboniferous limestone of the Monyash-Wirksworth area,' Quarterly Journal of the Geological Society, vol.113, 1959.
5. Butcher, N. D., and Ford, T. D., in litt 22.3.73 and verbal information.
6. Plan of the River Lathkill showing its connection with the Lathkill Dale Vein and Level. No date, c. 1836. J.H Rieuwerts private collection.
7. Derby Borough Library, Local History Library, Wyatt Papers.
8. Bar Coll., Section of Mandale Mine. No date, c. 1847-1848.
9. Estate Book, Over Haddon. (Section relative to mines and minerals.) In the possession of Mrs. W. Pearce of Over Haddon.
10. Thompson, S., In litt, 9.6.1972.
11. Grigor-Taylor, W. R., *'Notes on the Lathkill Dale Gold Mine.'* Bull PDMHS, vol.5, pt.1, April, 1972.

References

12 Hardy, W. and Houghton, W., *The Miners' Guide or The Compleat Miner*, 1810, p. xvii-xix.

13 *Victoria History of the County of Derby,* 1907, vol.2, p.327.

14 Nichols, J., *The History and Antiquities of the county of Leicester,* 1795-1815, vol.2, Pt.2, p.281.

15 Blanchard, I. S. W., *'The Duchy of Lancaster's Estates in Derbyshire, 1485-1540,'* Derbyshire Archaealogical Society Record Series, vol.3, 1967.

16 Melbourne Hall Estate Office. Documents relating to Over Haddon estate, seventeenth century, via the late Miss J. Wadsworth.

17 Kirkham. N., *'Lead Ore Tithe,'* Derbyshire Archaealogical Society. Local History Section, Supplement 9, 1965, p.10 and PRO Exchequer Depositions.

18 Woll Coll. Add mss., No.6678, f.1-14.

19 Dev Coll. Summary sheet of the Barmaster's receipts and expenditure, 1679.

20 *Articles of the Custom of the Mine.* Printed by Anne Ayrescough, 1721. Copy in Sheffield City Libraries.

22 Map of Over Haddon Estate, by Thomas Kirkland. nd, c. 1720-1727. In possession of Mrs. W. Pearce of Over Haddon.

23 Br-T Coll. L71, Ore measurement book, 1725-1734.

24 Br-T Coll. LP.19. Plan of the River Lathkill showing its connection with the Lathkill Dale Vein and Level. Staley, Surveyor, 1826. A very similar map is attached to a lease at Melbourne Hall (see 46).

25 Dev Coll. Lists of miners, ore burners and Barmote Court Jurymen, 1727-1728.

26 *Philosophical Transactions of the Royal Society,* Vol.43, 1744-45, p.266-267.

27 Bar Coll. Bundles of documents relating to trials in the Barmote Court. Mid-eighteenth century.

28 Dev Coll. as for 27, actually a complimentary series.

29 Bristol Archives Office, Council House, Bristol. City Archivist, in litt. 30.12.0.

30 Woll Coll. Add mss. No.6677, f.123-126.

31 Bar Coll. eighteenth century, Barmasters Books of entries, Private Liberties of the Duke of Rutland.

32 Dev Coll. Series of ore accounts,

References

lot and cope, eighteenth and nineteenth centuries. This is nearly a complete series of accounts of ore mined during the period in the Liberties leased to the Duke of Devonshire.

33 Derbyshire Record Office. Rieuwerts Collection, Kirkby Documents. Lists of Jurors for the Great Barmote Court for Youlgreave Liberty, 1770-1808.

34 Bag Coll. No.512, George Heywood's Cash Book, 1754-55.

35 Raistrick. A. *'The Mill Close Mine, Derbyshire, 1720-1780,'* Proceedings of University of Durham, Philosophical Society, vol. x, pt.1, 1938.

36 Northern Institute of Mining and Mechanical Engineers, Northumbrian Record Office, Newcastle. Extracts from Minute Books of London Lead Company.

37 Derby Mercury, 6.6.1777.

38 Br-T Coll. Miscellaneous documents, including L 362.

39 Verbal information from the late C. H. Millington of Monyash.

40 WHC, No.17.

41 WHC, No.23, Mandale Mine Reckoning Book, 1808-1837.

42 Derbyshire Record Office. Rieuwerts Collection, Kirkby Documents. Tithe ore collected in Youlgreave and Stanton Liberties, 1669.

43 Staffordshire Record Office. D1065, Pattinson papers.

44 Bag Coll. No.453 Barmote Court entries, Hucklow and other Liberties. 1812-1842.

45 Derby Borough Library, Local History Library. Taylor, Simpson and Moseley (Solicitors) Collection, No.364, Title to Lathkiln Dale Mine, 1782-1838.

46 Melbourne Hall Estate Office, Lease to Bateman and Alsop, 27.10.1827.

47 Dev Coll. Hartington and the Granges Liberty, Deputy Barmasters Book of entries, 1832.

48 Glover, S., *History and Gazeteer of the County of Derby*, 1829, vol.2, pt.1, p.387.

49 Robinson, P., *The Smiths of Chesterfield, 1957,* p.40.

50 *Mechanics' Magazine*, No.492, 12.1.1833.

There is another helpful drawing of the Dakeyne engine in 'The Draining of the Alport Lead Mines,' by Miss N. Kirkham, in Transactions of the Newcomen Society, vol. XXXIII. 1960-61. Miss Kirkham also comments, albeit very briefly, that the engine

References

may have been situated at the Lathkill Mine, p.88.

51 Dev Coll. Advert for the sinking of Lodge Shaft at Lathkill Dale Mine, 14.8.1830.

52 Lathkill Dale Mine Reckoning Book, 1834-1842 In possession of Mr. M. R. Cockerton.

53 Bagshaw, S., *History, Gazetteer and Directory of Derbyshire,* 1846, p.413. The Laxey wheel was not built until 1854 and the largest wheel in 1846 has not been identified.

54 The details of the miners and the financial breakdown and running costs of both Mandale Mine and Lathkill Dale Mine abstracted by the late Miss J. Wadsworth from the Census of 1841-61, the Mandale Mine Reckoning Book, the Lathkill Dale Mine Reckoning Book, and other sources.

55 Bag Coll. No.587 (III), Sale notice of machinery at Lathkiln Dale Mine, 16.9.1847.

56 Mort. J., the late Barmaster. In litt. 7.3.1960.

57 Bag Coll. No.549, contains printed prospectus of Mandale Mining Co. nd, c. 1839.

58 Br-T Doc, L 246, report to proprietors of Mandale Mine, May 1842.

59 Rieuwerts, J. H. *'Lathkill Dale; Its Mines and Miners,'* Bull PDMHS, vol.2, pt. 2, 1963, p.23.

60 Production figures from a variety of sources; Bar Coll. Br-T Coll. and Bag Coll.

61 Sheffield Independent, 12.1.1852. Notice of sale by auction, Mandale Mines.

62 Prospectus: Mycross Mines and Lathkilldale Level, 1885. In possession Mr. M. R. Cockerton.

63 Green. A. H. Memoirs of the Geological Survey, *'The Geology of the Carboniferous limestone ... of North Derbyshire,'* 2nd Edition, 1887, p.141.

64 Melbourne Estate Office. Miscellaneous mining correspondence, late nineteenth and early twentieth centuries, kindly supplied by Mr H Usher of Melbourne.

65 Tune R., A survey of Mandale Mine, Bull PDMHS, vol.4, pt. 1, 1969.

66 Butcher. N. In litt. 3.1.1967.

67 Rieuwerts. J. H., Ms notes of an exploration of Lathkilldale Sough, October 1959.

68 Bar Coll. O.S. 25in. to 1 mile, edition of 1899 with veins and soughs added by the Barmaster.

69 Ford, T.D. & Worley, N.E., Bull PDMHS, vol.6, pp 141-143, 1973

Subscribers' List

J Ambler, Wisborough Green, W Sussex
SP Allsop, Kirk Ireton, Derbys

R Bade, Morden, Surrey
C Bagshaw, Burton-upon-Trent
J Barnatt, Buxton
RA Bell, Bromborough, Wirral
RA Belson, Hellesdon, Norwich
A Bolton, Royston, Barnsley
Bookthrift, Ashbourne
C Bowden, Gt Hallingway, Herts
A Botham, Hanley, Stoke-on-Trent

RW Carrington, Bakewell
PJ Challis, Wirral
P Chandler, Tapton, Derbys
Ray & Caroline Cork, Ashbourne
DM Coulson, Mickleover, Derby
PR Cousins, Lichfield
D Crowther, Derby

D Dalrymple-Smith, Baslow, Derbys
P Deakin, Stoke-on-Trent

T Eades, Alstonefield, Ashbourne
C Eldred, Harston, Cambridge
AG Ellis, Macclesfield
English Nature, Over Haddon, Bakewell

Dr TD Ford, Oadby, Leics
JR Freeman, Allenton, Derby
ET Fretwell, Coundon, Co. Durham

D Gough, Newthorpe, Nottingham
Prof Gunn, University of Huddersfield

DJ Hale, Littleover, Derby
D Hart, Chesterfield
PR Hart, Carleton, Skipton
Hawkridge Bookshop, Castleton
John B Hicklin, Belper, Derbys
M Higgins, Edenthorpe, S Yorks
M Higginson, West Hallam, Derbys
David Hill, Hedon, Hull

RN Holden, Stockport

SD James, Wilford, Notts
Dr DP Jefferson, Old Dalby, Melton Mowbray

L Kirkham, Knypersley, Stoke-on-Trent
Dr JA Knight, Shirland, Derbyshire

C Lansdell, Norwich
M Luff, Coalville, Leics

K Makin, Walsden, Todmorden
DJ McCurdy, Golbourne, Warrington
Mike Moore, Moorebooks, Newport

D Noble, Stanton, Ashbourne

N O'Reilly, Wootton, Staffs
A Owen, Bolton, Lancs

Peak District Mining Museum, Matlock Bath
CLM Porter, Ashbourne
Jenny Potts, Hulland Ward

Mrs Rayner, Cressage, Shrewsbury
Mrs P Renshaw, Over Haddon, Bakewell
GO Roberts, Bakewell
Dr JA Robey, Mayfield, Ashbourne
Dr GE Roe, Bramhall, Cheshire

William A S Sarjeant, Saskatoon, Canada
RP Shaw, Athestone, Leics
K Smith, Bakewell
J Spencer, Didsbury, Manchester

W Taylor, Walton-on-Trent, Staffs

Revd ER Urquhart, Bakwell

G Warrington, Radcliffe-on-Trent
J Wood, Winster
T Wraxton, Lincoln
A Wright, Mutley, Plymouth

Index

A
Alsop, John 53, 60, 67, 93
Aqueduct 14, 16, 17, 29, 70, 87, 94
Aqueduct Level 29
Ashford Liberty 26, 27, 41

B
Bakewell Liberty 12, 14, 16, 37, 39, 91
Bateman, James 54, 55, 65, 74, 75, 94
Bateman, Thomas 53, 55, 60, 93
Bateman's House 13, 14, 22, 82
Black Marble 51, 80
Breeches Vein 12, 36
Bristol Company, The 32, 35, 38

C
Cantrell, George 49
Carter's Mill 20, 28, 38, 80, 87
Cornish engine 16, 89, 90, 94

D
Dakeyne brothers 55, 93
Dakeyne's engine 60, 61

E
Ecton Copper Mines 49

F
Fulling mill 26

G
Gank Hole Vein 12, 14, 19, 23, 43, 84, 94
Geology 6, 19, 108

Glossop, Richard 34, 35, 36, 84
Gold 23, 105
Granges Liberty 12, 35, 107
Greensward Rake 12, 39

H
Haigs Greave Vein 12
Highlow Pipe 39
Hillcarr Sough 40, 41, 43
Hubbadale Mine 38, 39

I
Inclined Plane 16, 79, 81

L
Lathkill Dale Sough 4, 32, 34, 37, 39, 43, 51, 82, 87, 91, 93, 94
Limonite 23, 84
London Lead Company 7, 33, 37, 38, 39, 93, 107

M
Mandale Vein 12, 29, 39, 43, 45, 50, 51
Milton Ironworks 94
Minerals 22, 99, 103, 105
Mycross Vein 12, 19

N
Nether Haddon Liberty 35

O
Over Haddon Field Rake 26
Over Haddon Liberty 14
Over Haddon Sough 28, 29, 91

P
Parker, Samuel 40
Pasture Rake 12, 14, 16, 29, 35, 43, 45, 47, 79

Q
Quo Warranto 16, 25, 91

R
Ridge Rake 12, 19, 34, 35, 36
Ringinglow Vein 12
Roberts, John 34
Roberts, Thomas 38, 39
Robinstye Mine 18, 20

S
Sideway Vein 12, 18, 19, 35, 84, 86
Smallpenny Vein 12, 19, 23, 37, 39, 93
Sough Mill 18, 19, 84

T
Taylor, John 40, 43, 49
Trevaskes, John 34
Tufa 32

U
Upper Haddon Liberty 32, 39, 43

W
Wager, William 47, 51, 105
Wakefield, John 49, 51, 105
Westgarth, William 38
Westgarth's engine 38
Winchester's Shaft 16, 26, 47, 50
Woodruff, Thomas 45
Wyatt, William 21

Y
Youlgreave Liberty 35, 49, 107

Other Mining titles available from Landmark

I would like to order copies of (tick box)

☐ *Lathkill Dale, Derbyshire: its Mines & Miners*
Paperback, 120pp, ISBN 1 901522 80 6, J. H. Rieuwerts, £6.95

☐ *Lead Mining in the Peak District*
Paperback, 208pp, ISBN 1 901522 15 6, Edited by T. D. Ford & J. H. Rieuwerts, £9.95

☐ *The Copper & Lead Mines around the Manifold Valley, Staffordshire*
Hardback, 256pp, ISBN 1 901522 77 6, L. Porter, £19.95

☐ *Derbyshire Blue John* Paperback, 112pp, ISBN 1 873775 19 9, T. D. Ford, £5.95

☐ *Glossary of Derbyshire Lead Mining Terms*
Hardback, 192pp, ISBN 0 904334 14 7, J. H. Rieuwerts, £14.95

☐ *Derbyshire Black Marble* Paperback, 96pp, £9.95

☐ *Derbys Lead Industry in 16th Century* Hardback, 338pp, ISBN 0 946324 10 7, D. Kiernan, £25

☐ *The Steam Engine in Industry: Mining & the Metal Trades*
Paperback, 128pp, ISBN 086190 544 X, G. Watkins, £7.95

☐ *Lead & Lead Mining in Derbyshire* Paperback, 89pp, ISBN 0 904334 09 0, A. H. Stokes, £5.95

☐ *Swaledale: Its Mines and Smelt Mills*
Mike Gill, available 2001, details on request

Photocopy this page for your order form

* Please add 10% of the publication price for post & package

I enclose my cheque for £.......... made payable to Landmark Publishing Ltd.

Or please debit my   Visa ☐   MasterCard ☐   Switch ☐

Card No: ☐☐☐☐ ☐☐☐☐ ☐☐☐☐ ☐☐☐☐

Expiry Date: ☐☐ ☐☐   Signature: ....

Name: ....

Address: ....

....   Post code: ....

Landmark Publishing Ltd
12 Compton (above David Neill), Ashbourne, Derbyshire DE6 IDA UK
Tel: 01335 347349   Fax: 01335 347303
e-mail: landmark@clara.net   www.landmarkpublishing.co.uk

LANDMARK COLLECTOR'S LIBRARY

LANDMARK COLLECTOR'S LIBRARY
THE SPIRIT OF
LEEK: 1
THE 20TH CENTURY IN PHOTOGRAPHS
Cathryn Walton & Lindsey Porter

LANDMARK COLLECTOR'S LIBRARY
THE SPIRIT OF
THE HIGH PEAK
THE 20TH CENTURY IN PHOTOGRAPHS
Mike Smith

LANDMARK COLLECTOR'S LIBRARY
THE SPIRIT OF
LICHFIELD
THE 20TH CENTURY IN PHOTOGRAPHS
H. Clayton & K. Simmons

LANDMARK COLLECTOR'S LIBRARY
STAFFORDSHIRE MOORLANDS
AND THE CHURNET VALLEY
• PHOTOGRAPHS OF DAYS GONE BY •
Lindsey Porter & Cathryn Walton

LANDMARK COLLECTOR'S LIBRARY
VICTORIAN TIMES
in and around
ASHBOURNE
PHOTOGRAPHS FROM THE 19TH CENTURY
Lindsey Porter

LANDMARK COLLECTOR'S LIBRARY
THE COPPER & LEAD MINES
around the
MANIFOLD VALLEY
North Staffordshire
Lindsey Porter & John Robey

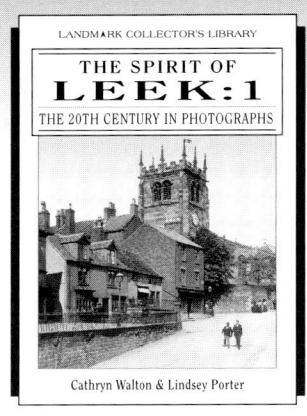

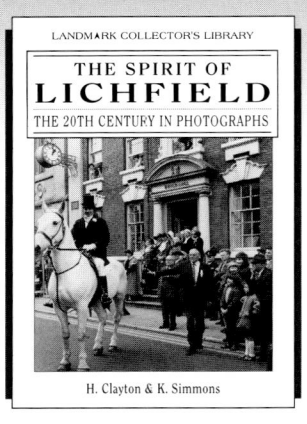

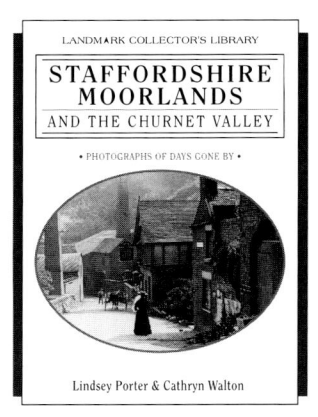

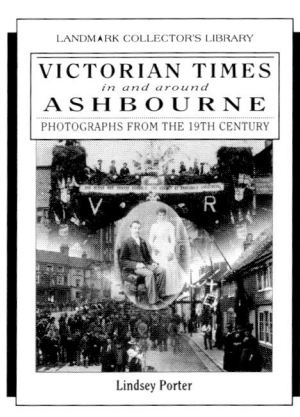

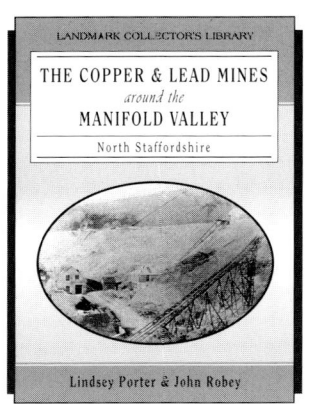

Hardback books on local history which you will enjoy having and dipping into time and again.

Produced to a high standard using art paper; carefully prepared and printed photographs; sewn bindings for extra life and attractively jacketed to enhance your book collection or to present as a gift.

Full details upon request

LANDMARK
Publishing Ltd

Waterloo House, 12 Compton, Ashbourne, Derbyshire DE6 1DA England
Tel 01335 347349 Fax 01335 347303
e-mail landmark@clara.net   www.landmarkpublishing.co.uk